MODULAR SYSTEM

JOSEPH WEST

ECOLOGY

Copyright ©

All rights reserved.
No part of this book may be
reproduced, stored in a retrieval
system or transmitted in any form
without the prior written permission
of the publisher.

PREFACE

Ecology is a rapidly developing branch of science. The major advances that are made, continuously affect our life on earth. Some of these important advances are included here.

The results of a recent survey on the attitudes to existing literature available to high school students showed that many were unhappy with the material used in teaching and learning. Those questioned identified a lack of the following; accompanying supplementary material to main text books, current information on new developments, clear figures and diagrams and insufficient attention to design and planning of experiments.

This book aims to improve the level of understanding of modern ecology by inclusion of the following; main texts, figures and illustrations, extensive questions, articles and experiments. Each topic is well illustrated with figures and graphs to ease understanding. Supplementary material in the form of posters, transparencies and cassettes will shortly be available. Profiles on common diseases are included in each chapter to inform, generate further interest and encourage students to explore the subject further. The 'Read me' articles supply up-to-date information on important issues related to each unit but outside the requirements of the current curriculum. It is the intention and hope of the authors that the contents of this book will help to bridge the current gap in the field of ecology at this level.

CONTENTS

I. The Scope of Ecology
- The scope of ecology 6
  - Terms used in ecology 8
  - Sciences related to ecology 9
  - Factors affecting the distribution of living things 10
- Test Your Knowledge 18
  - Choose The Correct Alternative 18

II. The Interactions of Life
- The interactions of life 20
  - Autotrophic organisms 20
  - Auto-heterotrophic organisms 21
  - Heterotrophic organisms 22
  - Food chain 26
  - Ecological pyramids 28
  - Energy flow 30
- Test Your Knowledge 32
  - Choose The Correct Alternative 32

III. Population Ecology
- Population Ecology 34
  - Features of population 35
- Test Your Knowledge 44
  - Choose The Correct Alternative 44

IV. Behavioral Ecology
- Behavioral Ecology 46
  - What is behavior and how it works .. 46
  - Types of behaviors 50
  - Society and its features 52
- Test Your Knowledge 56
  - Choose The Correct Alternative 56

V. Community and Ecosystem
- Community 58
  - Succession 59
- Ecosystem 61
  - What is bione 61
  - Biomes and climate 62
- Test Your Knowledge 68
  - Choose The Correct Alternative 68

VI. Biogeochemical Cycles
- Biogeochemical cycles 70
  - Water cycle 71
  - Carbon cycle 72
  - Nitrogen cycle 73
  - Phosphorus cycle 74
- Test Your Knowledge 76
  - Choose The Correct Alternative 76

VII. Human and the Biosphere
- Environmental Problems 78
  - Water pollution 79
  - Soil pollution 80
  - Air pollution 81
  - Noise pollution 85
  - Radiation 86
  - Measures against environmental pollution 87
  - Human and environment 88
  - Erosion and forests 93
  - Energy 93
  - Industrialization 94
- Test Your Knowledge 96
  - Choose The Correct Alternative 96

VIII. Appendix
- Scientific Measurement 98
- Glossary 100
- Reference 110

THE SCOPE OF ECOLOGY

Chapter 1

THE SCOPE OF ECOLOGY

Ecology is a branch of science that studies the interactions of living things with each other and with the environment.

Today there are over a billion organisms on earth. The complex relations of organisms with each other and with the environment is discussed and explained by ecology. Humans, a part of this system, live healthy, happy lives as the balance among living things and nonliving things is maintained.

Today environmental problems have increased due to developments in technology and industry, and because of the unwitting actions of people. For example, the overuse of chemical substances contaminates the water and causes slow-progressing diseases in humans; the inefficient use of soil depletes water sources and makes the soil arid. Such environmental problems increase the importance of ecology. For this reason ecology is taught as a science in schools, and private institutions keep people informed so that they will be more consciously aware of the environment.

For a thorough understanding of ecology, the relationships between organisms and the environment must be surveyed. Accordingly, the levels of organization are as follows: protoplasm - cells - tissues - organs - organ systems - organisms - population - community - ecosystem - biosphere - Earth - planets - solar system - galaxies - cosmos. Only the levels between organisms and biosphere are included in ecology.

> Ecology is the scientific study of the interactions between organisms and their environments ecology (from the Greek oikos, "home," and logos, "to study"). As an area of scientific study, many ecologists devise mathematical models by using sophisticated computer programs to develop models that predict the effects human activities will have on climate, and how climatic changes will affect ecosystems.

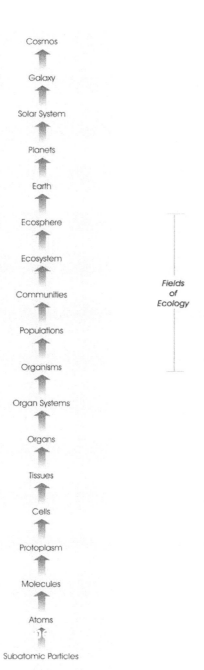

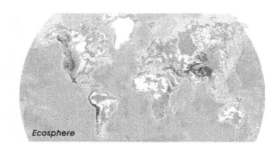
Ecosphere

Ecosystems
Communities
Populations
Organisms

Fields of Ecology

> For a thorough understanding of ecology, the relationships between organisms and the environment must be surveyed. Accordingly, the levels of organization are as follows: protoplasm - cells - tissues - organs - organ systems - organisms - population - community - ecosystem - biosphere - Earth - planets - solar system - galaxies - cosmos. Only the levels between organisms and biosphere are included in ecology.

Terms used in Ecology

- **Organismal (Individual) Ecology:** A branch of ecology that studies the relationship of an individual or individuals of a species to the environment.

- **Population ecology:** The next level of organization in ecology is the population, a group of individuals of the same species living in a particular geographic area.

- **Population:** The word "*population*" initially was used only for humans, but later was used for other organisms as well. Population is a group of individuals of the same species living in the same area. Population is the smallest unit of ecology. A population exists as long as it lives together with other populations and maintains its relations. In other words, one population is not self-sufficient.

- **Community:** A group of populations living together in the same area. With abiotic factors included, communities are self-sufficient.

- **Ecosystem:** A community together with the abiotic environment forms an ecosystem. Environment is the place where an organism lives. The environment is made up of abiotic and biotic factors. e.g., air, light, water, humans, other organisms and all nonliving things form the environment.

- **Biosphere:** All of the places where organisms can live, from the bottom of the ocean to an altitude of 10,000 m.

- **Habitat:** The natural environment or place where an organism, population, or species lives. It is shortly address of the organism. For example habitat of paramecium is fresh water and habitat of certain kind of ants is trees in the forest.

- **Niche:** The function of a particular species in the environment. All the processes of the individual are included in this term.

- **Biome:** The geographical area of the environment that an organism needs to live. Biome can be thought of as the place where the community lives.

- **Biomass:** the dry weight of organic matter comprising a group of organisms in a particular habitat.

- **Flora:** The plant populations living in a particular environment.

- **Fauna:** The animal populations living in a particular environment

> Interaction is a key idea in ecology. No organism is completely self-sufficient. Organisms depend upon other organisms and upon the environment for survival.
>
> For example autotrophs produce food and oxygen and heterotrophs produce carbon dioxide, which is needed for autotrophs to produce food.

Sciences related to Ecology

Certain branches of science are closely related to ecology. The science that studies the distribution of plants and animals is called biogeography; the science that studies abiotic and biotic factors of fresh water is limnology; the branch that studies the biotic and physical conditions of marine ecosystems is called oceanography; the ecology of radiation is called radiology; and the ecology of space is called space ecology.

The progresse made in the field of ecology has led to the formation of new branches of ecology. These are shown in the following figure.

It was initially stated that ecology is the study of organisms and their relationship to the environment. To understand ecology better, first environmental factors that affect living things, and then their relations with each other and their environment will be studied.

Living things exist in relationship to biotic and abiotic factors. Abiotic factors are studied in two sections: climate and soil. Biotic factors are living things. Living things are placed in 3 groups: producers, consumers and saprophytes (*Figure-1.1*).

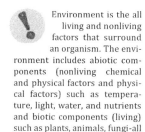

Environment is the all living and nonliving factors that surround an organism. The environment includes abiotic components (nonliving chemical and physical factors and physical factors) such as temperature, light, water, and nutrients and biotic components (living) such as plants, animals, fungi-all the other organisms.

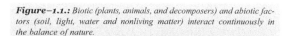

Figure–1.1.: *Biotic (plants, animals, and decomposers) and abiotic factors (soil, light, water and nonliving matter) interact continuously in the balance of nature.*

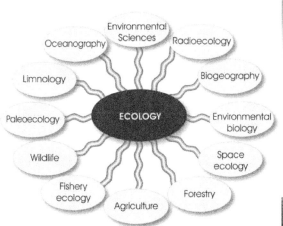

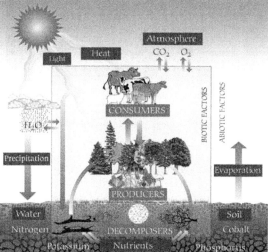

> Environmental temperature is an important factor in the distribution of organisms because of its effect on biological processes and the inability of most organisms to regulate body temperature precisely.

Factors affecting the distribution of living things

Environment is the all living and nonliving factors that surround an organism. The environment includes abiotic components (nonliving chemical and physical factors and physical factors) such as temperature, light, water, and nutrients and biotic components (living) such as plants, animals, fungi-all the other organisms.

Abiotic factors of the biosphere

Abiotic factors are important determinants of the distribution of organisms in the biosphere.

Climatic Factors:

- Light
- Temperature
- Water

Soil Factors:

- Structure of soil
- Minerals and salts
- pH of soil

Climatic Factors

The variety of living things on earth is affected and determined by sunlight, temperature, pressure, moisture, and air movements, which are altogether called climatic factors. When the weather of an area is mentioned, it means all the above factors at that time or in that year.

The science of climate is meteorology. The climate of an area refers to the annual average of light, temperature, rainfall, and air movements in the area over many years.

A related science is climatology. Weather is studied by meteorologists, who determine the average values of atmospheric characteristics for an area. These values determine the living area of an organism. For example, areas with heavy rainfall and temperate climate are suitable for a jungle ecosystem and the dependent jungle animal populations. At the same time, climatic conditions are the main factors that determine the distribution of living things and habitat formation. The factors that cause the formation and change of climate are temperature, light and water.

Light

As explained before, the energy source for all organisms in nature is light. The natural source of light is the sun (*Figure-1.2*). From an ecological point of view, the intensity and duration of light is important. The amount of light energy at a certain time and place is related to the angle of incidence of light rays. As the angle of incidence increases, light rays pass through a thicker layer of atmosphere than rays that strike the atmosphere at a right angle and more light is captured in the atmosphere. Thus less light reaches the earth's surface.

> All organisms must live within a certain range of temperatures. In general, warm-blooded animals are active in a wide range of temperatures. Cold-blooded animals are active in narrow range of temperatures. Animals generally cannot survive in temperatures that exceed 52°C. Some types of algae survive in hot springs, where temperatures may be 73°C or even higher. Some types of algae live in artic ecosystems. Extraordinary adaptations enable some organisms to live outside this temperature range.

As the angle of incidence increases, the incoming light rays are spread over a greater surface and the amount of light per surface unit is proportionally less. The reproduction, migration, and pigmentation of various organisms are all affected by light, as is respiration, especially of those organisms, living in wet environments, resulting in decreased oxygen consumption.

Light is essential for photosynthesis. Light intensity and photosynthetic rate are directly proportional. In both shade and sun, light intensity increases the rate of photosynthesis. If the food produced at the maximum level of photosynthesis is higher than the amount consumed in low light, the plant will store the excess. The stored food is eaten by animals.

In tropical forests, long day plants form the canopy. (e.g., acacia, willow).These are broad-leaved, large-celled and large-stomated plants. Below the canopy are understory plants. The plants of this layer include banana, dogrose, and ivy. Beneath there are herbaceous plants such as ferns, horsetail and some grasses. (*Figure-1.3*).

Shade density increases with forest density. The development of the canopy and forest regeneration are proportional to the shade tolerance of the young trees. If the plants are resistant to shade, then forest regeneration will be easy; if they are not, regeneration will be difficult and the forest may undergo a new formation as dominant species are replaced. Light is also a factor at work in aquatic ecosystems. Light, depending on the transparency of the water, can penetrate up to 200m below the water surface.

Light is used as an energy source by aquatic plants and effects pigment production, the exoskeleton and shell, and the formation of some other structures in varioius animals.

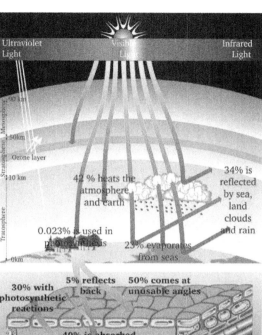

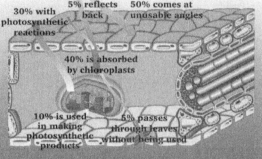

Figure–1.2.: *The source of energy in ecosystems is sunlight. Only a small amount of sunlight can penetrate the atmosphere and reach the earth. Some of this light is absorbed by plants. In the figure above, a cross-section of a leaf and the storing of light enery in organic molecules in chloroplasts are demonstrated.*

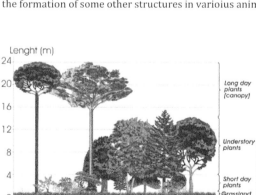

Figure–1.3.: *Forests have four regions, namely canopy, understory, shrubs and grasses. The factor that determines the formation of these regions is light.*

READ ME!
Heat-Temperature-Freezing

Heat and temperature describe two physical features that are closely related to each other. They are frequently confused. It is important to understand that heat and temperature are quite different.

Heat is the potential energy present in the mass of an object. It is the energy that keeps the molecules inside an object in motion. It is also called internal energy. Heat is measured in units called calories. One calorie is the energy needed to raise the temperature of 1 gram of water 1 degree Centigrade (from 14.5 to 15.5 C). Another unit of heat measurement is the joule (J), where 1 calorie= 4.184 J

Temperature is a measurement of hotness or coldness. Heat is a form of energy, whereas temperature is a measurement of the intensity of heat. Heat always flows without outside help from warm objects to cool ones. For example, a cup of hot coffee cools because heat flows from the hotter coffee to the cooler surroundings.

Temperature is measured with the Centigrade(C) or Fahrenheit (F) scale. The main source of heat is the sun. The amount of heat energy coming from the sun in the form of light depends on various factors. The amount of light reaching our planet is highest at the equator and between the tropics; it is lowest at the poles (the effect of latitude). Accordingly the temperature is higher near the equator and lower at the poles. In the same way, at higher altitudes the weather is usually cold. The thawing of frozen soil causes the water below to rise, evaporate and mix into the air. If the temperature falls below the freezing point, a thin layer of topsoil freezes. The bottom surface of this frozen layer attracts water from the soil, forming a thin layer of ice. As the layer of ice increases, the frozen soil swells. This swelling is called **frost heave**. In this way, a cycle develops of swelling at night and melting during the day. This in turn causes the roots of plants to be pulled up more every day. If this cycle continues, the roots of young plants may be pulled up 8-10 cm, and the plants will die. Most of the time when the plants are pulled up, their roots are torn, causing death.

Temperature

Sunlight transports energy from the sun to the earth. The light that passes through the ozone layer energizes molecules in the atmosphere and, consequently, supplies heat energy (temperature) to the living and nonliving things on earth. This process is very important for living things because all life on earth needs heat to survive. The source of this heat is sunlight.

Effect of Temperature on Plants

The temperature of a plant is directly related to the temperature of the environment. Generally, the temperature of plant roots depends on soil temperature. The parts of the plant above ground, when they absorb sunlight, are a few degrees higher than air temperature. During transpiration the temperature of these parts is lower than the air.

The effects of low temperatures on plants is not seen everywhere and every time to the same degree. When the temperature declines slowly, plants get rid of excess water and can accommodate to the low temperature to a certain level.

In winter, especially at night, when air temperature declines substantially, the temperature of the plant stem declines as well. As a result, some shrinking or wrinkling may appear on the stem. Though at low temperatures the bark of trees creases due to heat loss, the inner structures of the stem experience no shrinkage because they are warmer.

Tree bark can't resist this strong creasing and tension, and so it splits longitudinally. In winter, if the soil is cold and frozen, but above ground the air temperature is high, plants can't replenish the water lost in transpiration with water from the soil through their roots because low temperatures increase the density of water. As a result, plants suffer from what is called physiological drought.

If the temperature rises above the plant's optimal temperature, it experiences a stagnation phase. If the temperature rise continues, the plant can't compensate with water from the soil that water that is lost in transpiration. Then, beginning with the leaves, the green parts become yellow. Later the protoplasm clots and the plant dies.

Effect of temperature on animals

Animals are placed into two groups according to the relationship between their body temperature and the air temperature: **poikilothermal animals** (without constant body temperature) and **homoiothermal animals** (with constant body temperature).

Invertebrates, fish, frogs and reptiles are poikilothermal animals. These animals have a body temperature close to the ambient temperature. Homoiothermal animals, though they have a constant body temperature independent of the ambient temperature, may experience slight temperature changes due to external conditions. Temperature affects the development, reproduction and metabolism of organisms. Extreme temperature changes cause death.

Temperature range and survival

Some organisms that have a metabolism that normally functions between 0-50 C can live below 0 C or above 50 C as well. Some bacteria can survive temperatures of 90 C, some mollusks live in water at 46-48 C, and some fishes can live in water with a temperature above 40 C.

Organisms can decrease their metabolic rate to adjust their body temperature to the ambient temperature. As in the examples given above, the tolerance of species to temperature varies. For every species there is an upper and lower limit. (*Figure-1.4*).

However there are minimum and maximum temperatures at which organisms decrease or increase their activity to survive. Organisms normally seek the optimum temperature. Organisms are divided into two groups according to tolerance to temperature: Those that can tolerate slight changes and those that can tolerate broad temperature changes. (*Figure-1.5*).

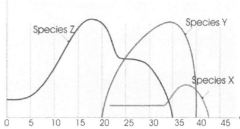

Figure–1.4.: *The above graph shows the populations of insect species X, Y, and Z in an ecosystem at different temperatures. The following observations can be made: The optimum temperature intervals of species X and Y are similar. Species Z is the most resistant to cold. Species Y and Z reach their maximum populatioin size at different temperatures. Of these species, the most resistant to high temperature is species X.*

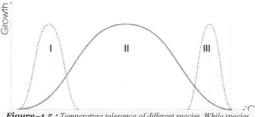

Figure–1.5.: *Temperature tolerance of different species. While species I and III adapt to a narrow range of temperatures, species II is adapted to wide range of temperature.*

Water

Water precipitates from the atmosphere in the form of rain, snow, and hail. Annual rainfall varies in different parts of the world. Latitude, large bodies of water, mountains and wind affect precipitation. Organisms can't live without water and there is no substitute.

Water vapor enters the atmosphere through evaporation, where it absorbs most of the light reflected from the earth's surface, which in turn prevents the excess warming and cooling of the earth. As humidity decreases, evaporation and transpiration rates increase. Plants need water and exchange it with the atmosphere. Plants need at least 65% humidity to maintain this balance.

Moisture is another factor that determines climate. Moisture includes both water that falls from clouds, or precipitation, and water vapor in the air, or humidity

Effect of water on plants

Plants are placed in three groups according to their water needs or structural differences arising from the amount of water.

- Hydrophytes (grow where water is always available)
- Mesophytes (grow where water availability is intermittent)
- Xerophytes (grow where water is scarce most of the time)

Figure–1.6.: Aquatic plants like water lily, a hydrophate, have features adapted to their environment. Wide leaves and the placement of stomata on the upper surface of the leaves ease the process of evaporation.

Hydrophytes: Hydrophytes live in water and therefore have no problem with transpiration. Roots may be in or out of the water. Stomata are present on the upper surface of the leaves and the leaves are covered with a thin layer of cuticle. They don't have any water related problems. Examples include water lily and elodea. (*Figure-1.6*).

Figure–1.7.: These plants have zigzagged leaves with a wide surface. This feature enables greater light absorption, a high rate of photosynthesis, and faster transpiration.

Mesophytes: These plants are adapted to live in places where water supply is intermittent. Cuticles are of intermediate thickness; stomata are present on both upper and lower surfaces of the leaves. (*Figure-1.7*).

Xerophytes: Xerophytes are adapted to arid conditions like deserts and sand dunes. They have a very extensive root system. Some xerophytes store water inside their bodies after rainfall. (*Figure-1.8*).

Their cylindrical and spherical shapes produce a small surface area, so they lose less water through transpiration. They also have a thick cuticle, and stomata that open at night instead of during the day.

Generally, xerophytes have properties that decrease transpiration. Their leaves are thick, needle-like and small-celled. Stomata may be covered with hair-like structures or protected folds of the leaf, or they may be embedded beneath the leaf epidermis. Moreover, cells have high osmotic pressure.

Figure–1.8.: Desert and arid region plants. The leaves of these plants are either spiny or needle like. This makes for a smaller surface area for transpiration, an adaptation for life in arid regions.

Plants living in salty environments also have the properties of xerophytes, but it can't be said that these plants can live in arid places as well.

Effect of water on animals

Animals obtain water with active processes. Their means of water acquisition are:

- Ingesting water directly through the digestive system.
- Wet-skinned animals (living in soil, mud, and sand) absorb water through the skin.
- Water present in ingested foods.
- Water released from the food in metabolism.

To save water, animals have mechanisms like those of plants. For example, skin minimizes water loss. Most organisms lose water and salt from their sweat glands to maintain body temperature. The body maintains water balance by taking in enough water to compensate for the excreted water.

The animal respiratory system has an important role in conserving body water. Since fish live in water, they don't have a problem. In terrestrial organisms though, the respiratory system is inside the body (lung, trachea). A small amount of water is lost by insects and terrestrial mollusks (snails). Closing of respiratory holes in arid times minimizes water loss. Animal excretory systems also play an important role. Aquatic organisms excrete ammonia, a very toxic substance, with substantial water. Terrestrial organisms convert ammonia into less toxic urea. Reptiles and insects living in arid areas convert ammonia to uric acid and excrete it with little water. Animals' metabolic reactions provide some water. Termites eat dry wood continuously. When termites digest wood and metabolize it, water released as a result of metabolism quenches their thirst. Desert camels can survive without drinking water for 11 days by using metabolic water produced in the catabolism of fat in the hump. In the same way, hibernating animals, like bears, and migrating birds obtain water as a result of the metabolism of fats stored in the body. Antelope and some rodents can survive on guttation water.

Humidity

Humidity is very important for storing crops. If the storehouse has high humidity, then the water content of the crop will increase and the crop will rot. Appropriate humidity levels for storehouses are 38% at 10 C, 45% at 16 C, 50% at 21C, and 60% at 33C. Like humidity, fog is also important, especially in arid climates. In such places when fog touches the plant or soil, some water condenses on the surface because of its lower temperature. In plants, the cuticle and epidermis layers absorb water and this causes an increase in turgor pressure.

In other words, fog affects plant growth positively by moistening the soil and increasing turgor pressure. There may be no rain in the desert for a few years, but some plant species survive by water absorption. The best example of this situation is the flora of the Atacama Desert in Chile.

However, if there is excess humidity, or during long periods of foggy or cloudy weather, plants may get some diseases, because some fungi grow very fast under these conditions and spread, causing a decrease in plant production.

READ ME!
The Bergman, Allen and Gloger Rules

Bergman's Rule: Temperature is a factor that determines the size of animals. Homoiothermal (warm blooded) animals living in northern latitudes tend to be bigger than their relatives in the hotter regions. Bergman's rule states that populations in colder climates (higher latitudes) have larger bodies than populations in warmer climates (lower latitudes).

With a larger body, the surface/volume ratio decreases. Large-bodied animals have a relatively smaller surface area and preserve internal heat more efficiently. This is an important adaptation. For example, the size of penguin species increases from South America to Antarctica. The size of bears and hares increases from south to north in their ranges.

But there are examples of the opposite as well. In the poikilothermal (cold-blooded) animals like frogs and reptiles the reverse is seen. These animals have smaller bodies in cold regions than in hot regions. In other words, in poikilotherms, as air temperature decreases body size decreases as well.

Allen's Rule: Allen's rule states that animals in colder climates generally have shorter extremities (beaks, wings, ears, feet) than those populations in warmer climates. In hares and foxes, organs like the ears are smaller from southern to northern regions. In hot regions, a large surface area is a means of transpiration and cooling for the animal. Allen's observations have been demonstrated experimentally. Laboratory mice grown at a temperature of 31-33.5 C have longer tails than those grown at a temperature of 15.5-20 C.

Gloger`s Rule: Gloger's Rule states that populations in warmer and more humid climates have darker coloration than those in cooler or drier climates. In the northern hemisphere birds and mammals have lighter colors from the Equator north; and have darker colors from north to south in the southern hemisphere. In the formation of colors environmental and genetic factors interact.

Soil Factors

Soil is another factor with which living things are continuous interacting, directly or indirectly. Soil structure, pH, mineral and salt content affect the organism in different ways.

Soil structure

When you glance at the soil, it seems that it is nonliving, but actually it is full of billions of organisms. Soil is very suitable to life for bacteria, fungi, viruses, algae, and protozoans.

Microorganisms are especially abundant in soil rich in organic wastes. The dominance of a microorganism in a certain area is determined by environmental conditions. For example, in fall yeast cells are more abundant in soil that is covered with ripe fruit. These cells produce alcohol from glucose

After yeast cells, Acetobacteria are second in dominance. After rain, low-lying areas become swampy, which prevents the diffusion of gases into the soil. As a result, aerobic bacteria are replaced by anaerobic bacteria.

For instance, the process of nitrification carried out by aerobic bacteria like nitrosomonas and nitrobacteria is replaced by a denitrification process carried out by anaerobic bacteria.

The number and variety of soil bacteria are greater than that of all other soil organisms. These bacteria may be autotrophic or heterotrophic, aerobic or anaerobic.

Soil, water, air, organic and inorganic molecules are very important for plant growth. The ratios of these 4 groups in the soil are as follows:

- Minerals (Ca, Mg, P, N): 45%
- Organic molecules (plant and animal residues): 5%
- Air: 25%
- Water (soil water with dissolved salts): 25%

Minerals and Salts

Organisms contain very important and vital minerals. The most important ones are N, P, K, Ca, S, Fe and Mg. Deficiency of these causes serious problems in living things. For example, Ca is an element used by all organisms. Calcium is a constituent of animal endo- and exoskeletons, and is necessary for muscle contraction and blood clotting. Moreover, it has a role in the adjustment of soil pH and in the density of soil water. Magnesium is present in the chlorophyll and also works as a cofactor of enzymes in DNA replication. (*Figure-1.9*).

Some elements may be present in sufficient amounts for the survival of organisms. The measurement of the amounts of these elements is very difficult, but can be determined by the isotope method. If one of these elements is missing, pathological symptoms are seen in plants, animals and humans.

The essential minerals and elements for living things are Fe, Mn, Zn, B, Na, Mg, Cl, and Vanadium. Every organism needs these elements in different quantities. At present, certain important minerals, especially N, P and S are mixed into the soil as inorganic fertilizers to meet the requirements of plants. If excess fertilizer (minerals) is applied to the soil, plants can't absorb water because of the increased density of soil (physiological drought). Consequently, plants get yellow and die.

Soil pH and plant relations

Soil pH means its degree of acidity or alkalinity. This depends on the amounts of hydrogen (H) and hydroxyl (OH) ions. Cultivated plants mostly need soil with pH 6.7-7.0.

The main reason an increase in soil acidity is a decreased level of Ca. For this reason, highly acidic soils are treated with lime (CaO) to decrease acidity. Spoiled soil is a result of an accumulation of the minerals Ca, Mg, Na and salts like Cl_2^{-1}, CO_3^{-2}, HCO_3^- and SO_4^{-2} ions. An overabundance of these salts will increase the concentration of soil water and makes the absorption of water and minerals from the soil difficult. (physiological drought).

Soil is another abiotic factor of ecosystems. Soil is important to plants as a source of minerals and as a material in which to anchor their roots. Many animals also depend on soil for a place to live and for food. The physical structure, pH and mineral composition of rocks and soil.

They limit the distribution of plants and animals that feed upon them.

Figure–1.9.: *The deficiency of the essential minerals in the environment influences the metabolism of plants. As seen in the figure, the deficieny of a certain mineral affects plant growth differently.*

CHOOSE THE CORRECT ALTERNATIVE

1. Which one of the following is a biotic component of an environment?

   A. Soil   B. Light   C. water
   D. Temperature   E. Pine tree

2. Which one of the following is the most inclusive level of organization in world?

   A. community   B. ecosystem   C. cell
   D. biosphere   E. population

3. Which of the following is the most important climatic factors affecting the distribution of organisms?

   A. Water and wind
   B. Temperature and sunlight
   C. Predators and parasites
   D. Wind and sunlight
   E. Prey and predators

4. The sum of all Earth's ecosystems is called the _____.

   A. biome
   B. atmosphere
   C. biosphere
   D. hydrosphere
   E. ozone

5. Which of the following branch of science is closely related to ecology?

   A. Maths
   B. Chemistry
   C. Cytology
   D. Histology
   E. Environmental biology

6. Which of the following is a mesophyte plant?

   A. Water lily   B. Cacti   C. Elodea
   D. Oak tree   E. All of the above

7. Which of the following animal is a homoeothermic?

   A. Cat   B. Frog   C. Snake
   D. Eagle   E. All of the above

8. Which of the following is an abiotic component of environment?

   A. Lion population
   B. Pine tree population
   C. Prey and predators
   D. Soil and elephants
   E. Light and temperature

9. What is habitat?

   A. A place in where the individuals of population live
   B. The function of a particular species
   C. A group of populations living together
   D. A plant population living in the sea
   E. An animal population living in the forest

10. What of the following describes flora?

    A. A place in where the individuals of population live
    B. The function of a particular species
    C. A group of populations living together
    D. A plant population living in the sea
    E. An animal population living in the forest

THE INTERACTIONS OF LIFE

Chapter 2

THE INTERACTIONS OF LIFE

One of the properties that distinguishes living things from nonliving things is their nutrition. Organisms feed and acquire materials necessary for energy production, regulation and assembly. Organisms are classified according to their feeding styles.

Autotrophic organisms

Autotrophs produce their own food from inorganic substances. Autotrophs are either photosynthetic or chemosynthetic according to the energy used.

Photosynthetic autotrophs: These organisms produce organic molecules from inorganic molecules using sunlight energy (photosynthesis). Green plants, algae, and blue-green bacteria are photosynthetic autotrophs. Some bacteria use hydrogen sulfide (H_2S) or hydrogen (H) instead of water. The bacteria using these don't release O_2.

Chemosynthetic autotrophs: Some bacteria oxidize inorganic substances and release energy. From this energy ATP is synthesized. ATP is used in the production of organic substances from inorganic ones. Since chemicals are used in place of light, this food synthesis is called chemosynthesis. Examples of chemosynthetic bacteria are nitrite and nitrate bacteria.

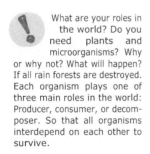

What are your roles in the world? Do you need plants and microorganisms? Why or why not? What will happen? If all rain forests are destroyed. Each organism plays one of three main roles in the world: Producer, consumer, or decomposer. So that all organisms interdepend on each other to survive.

Auto-heterotrophic organisms

Some unicellulars (e.g. Euglena) carry chloroplast and practice photosynthesis. These organisms carry out photosynthesis in the presence of light (autotrophs), while at night they obtain food from their surroundings (heterotrophs).

Insectivorous plants have chloroplasts like green plants and carry out photosynthesis. At the same time, since these plants live in nitrogen-deficient soil, they obtain nitrogen by eating insects. (*Figure-2.1*). They secrete enzymes to digest insect proteins. The amino acids released are absorbed into the cells and used in metabolism. Examples include dionea, drosera and nepentes plants.

> Carnivorous plants are exactly that, they are meat eaters or more correctly insect eaters hence their true title Insectivorous plants (Carnivorous sounds far more interesting). These plants lives in areas where there is a lack of nutrients in the soil or very poor conditions where roots do not survive well. The plants have adapted their leaves into insect catchers that can hold down an insect and digest the internal juices. Insect eaters are therefore highly advances plants quite different from other plant types.

Figure–2.1.: Some plants living in nitrogen-deficient soils obtain nitrogen by ingesting insects. These plants are called insectivorous plants. Some examples are dionea, drosera and nepentes that attract insects with their color and aroma. The trapped insects are digested with enzymes secreted by the plant and then metabolized.

Heterotrophic organisms

Animals, fungi, some bacteria and protists can't synthesize their own food and get it from other organisms or decaying matter. Heterotrophs have different types of nutrition according to their habitat and food type used.

Holozoic nutrition (Heterotrophic)

This is the form of nutrition used by most animals and involves the ingestion of complex food, which is broken down into simpler molecules before being absorbed.

- **Carnivores (meat eaters):** They eat only meat. Examples are lion, tiger, wolf and etc.

- **Herbivores (plant eaters):** They eat only plant. Examples are sheep, gazelle cows and etc.

- **Omnivores (plant and meat eaters):** They eat both plant and meat. Examples, humans, monkeys, birds and etc.

The digestive systems of these organisms vary according to the type of food. For example, herbivores have well-developed molar teeth, 4-chambered stomachs, and long intestines, because the digestion of grass is difficult.

Carnivores have well-developed incisor and canine teeth, single-lobed stomachs, and shorter intestines.

Omnivores have the properties of both moderately.

> All organisms need energy to live and complete their life cycle. The main source of energy is the radiant energy from the sun but it is unusable by all organisms.

Symbiotic Nutrition (living together)

Some organisms live in close relationship. There are types of this relationship.

Commensalism

If one organism benefits and the other is neither harmed nor helped, the relationship is called commensalism. (*Figure-2.2*).

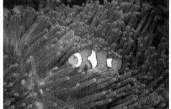

Figure–2.2.: *Commensalism in the sea. Clownfishes (Amphiprion perideraion) often form symbiotic association with sea anemones, gaining protection by remaining among their tentacles and gleaning scraps from their food.*

The helping organism is called commensal. For instance, small fish (Echeneis) attach to sharks and live with them. These small fish feed on the residue of the shark's prey. Here, while the small fish benefit, the shark (commensal) is not affected.

Mutualism (Mutual Benefit)

In this type of relationship both organisms benefit. Lichens are a typical example. Lichens are composed of fungi and green algae. Fungi protect the algae and provide them with water and CO_2. Green algae supply the fungi with food and O_2.

Another example is the relationship between the rhizobium bacteria (Rhizobium leguminosarum) and legume plants. These bacteria live in the root nodules of legume plants (*Figure-2.3*). Saprophytic rhizobium bacteria live in the soil and when they encounter the roots of legume plants, they enter the root hairs and pass to the cortex cells, where they reproduce using the food and enzymes of the plant. Host cells activated by the bacteria multiply quickly and form pocket-like bacteria-containing nodules. Here the plant gets the advantage of atmospheric nitrogen, which is fixed by the bacteria. The plant provides the bacteria with shelter and the products of photosynthesis

> Biotic (plants, animals, and decomposers) and abiotic factors (soil, light, water and nonliving matter) interact continuously in the balance of nature.

Figure–2.3.: *Rhizobium bacteria living in root nodules of legume plants are a good example of mutualism. The bacteria supply the plant with nitrogen, and the plant supplies the bacteria with the products of photosynthesis.*

Figure–2.4.: Flea and louse, external parasites, are very dangerous to humans. These organisms don't have a well-developed digestive system and suck blood for nutrition.

Parasitism

Parasitism is the symbiotic relationship in which one member (parasite) benefits and the other (host) is adversely affected.

Parasites have well-developed sense and grasping organs and reproduce quickly. On the contrary, their enzyme and digestive systems are not well-developed. Parasites live in or on the host. They suck liquid nutrients from the host.

Parasitic animals Parasitic animals may be internal or external. Both groups contain different organisms. Internal parasites don't have digestive systems and live in places where digested food is available. External parasites can partially digest food. Examples of external parasites are lice, fleas and bedbugs (*Figure-2.4*). Examples of internal parasites are plasmodium, tapeworm, taenia, roundworms and flukes.

Parasitic plants: Some plant species live on other plants and obtain organic or inorganic substances from them. Such plants are of two types: half-parasitic and full-parasitic.

Full-Parasitic plants: The organs of these plants have certain peculiarities. Leaves are small, with little or no chlorophyll, weakened xylems and, in some cases, roots disappear.

The absence of photosynthesis is compensated for by the development of sucking organs called haustorium. (*Figure-2.5*).

Figure–2.5.: Orobanchaceae (broom-rapes), a full parasitic plant, does not carry out photosynthesis but obtains nutrients through haustoria from the host plant.

Parasitic plants anchor their haustoria to the vascular tissue of the host plant and absorb and use the food produced by the host. In the same way they obtain the water necessary for transpiration.

Parasitic plants cause enormous harm to cultivated plant. Examples include Orobanchaceae (broom-rapes) and Cuscutaceae.

Half-Parasitic plants: These plants anchor their haustoria into the xylem of the host plant, absorbing water and minerals which they use to produce organic substances.

They have chlorophyll and also carry out photosynthesis. Mistletoe, a half parasite, lives on trees such as apple and pear. (*Figure-2.6*).

Pathogens: Many bacteria and fungi live parasitically on higher plants and animals and cause disease. In other words such parasites are at the same time pathogens.

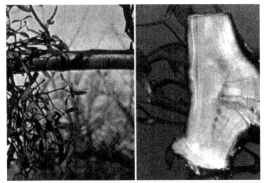

Figure–2.6.: Half parasitic plants, mistletoe obtain only water and minerals from the host plant.

Parasites that can't survive unless they are on a host organism are called **obligatory parasites.** The bacteria that causes diphteria is such an organism, unable to survive outside the human body.

The bacteria that cause cholera and tetanus can live in soil or water in a dormant condition. When they find suitable conditions they become parasitic and pathogenic. Viruses are also obligatory parasites.

Saprophytic nutrition (decomposers)

Saprophytic nutrition is a type of heterotrophy. Some bacteria and fungi feed on and digest organic substances in decaying animal and plant remains. These organisms, also called decomposers, have a well-developed digestive system.

They practice extracellular digestion and convert organic substances into inorganic ones. In this way they clean the environment and contribute to the nitrogen cycle.

Bacteria and fungi that get food from the nitrogenous organic compounds of dead plants and animals cause decomposition and putrefaction, and enable matter to cycle in nature.

Preteus vulgaris is one of the saprophytic bacteria that causes putrefaction (*Figure-2.7*).

Figure–2.7.: *Some bacteria and fungi, living as saprophytes, decompose dead plants and animals, making their nutrients available and participating in the cycling of matter.*

Parasitic plants

For a good understanding of the problems these parasitic plants can cause, it is necessary to give some background information about the biology of parasitic plants in general and, more specific, about the various stages of the life cycle and the interaction with the host plant during these stages. Interactions between organisms in nature occur frequently. These interactions can be symbiotic. This term includes both mutualistic symbiosis (all the organisms involved benefit from the association) and parasitic symbiosis (only one of the organisms involved benefits to the detriment of the other organism of the interaction). Associations involving higher plants can be found on two levels. The majority comprises the interaction between higher plants and mycorrhizas or nitrogen-fixing bacteria. These associations are mutualistic symbiotic in nature. On the contrary, interactions between higher plants are usually parasitic, involving a non-parasitic host and a hemi- or holoparasitic plant.

It has been estimated that about 1% of all flowering plants, roughly 3000 species, is parasitic. They form a close connection with the vascular system of the host through a so-called haustorium and are at least partially dependent on the host for their supply of water, nutrients and organic solutes. For a good understanding it is necessary to define specific terminology. Parasitic angiosperms may be classified as either root parasites (60%) or stem parasites (40%), depending on whether the haustorium is below or above soil surface.

They also may be divided into groups with regard to the presence (hemi-parasites) or absence (holo-parasites) of chlorophyll. Approximately 20% of all parasites is holo-parasitic, the remainder being hemi-parasitic. All chlorophyll-lacking species are obligate parasites, meaning they cannot establish and develop independently. Other parasitic plants are facultative parasites; they can establish and grow independently but in field situations always meet a wide variety of hosts and behave heterotrophic.

Food chain

A food chain consists of producers, consumers and decomposers. All organisms need energy to live and complete their life cycle. The main source of energy is the radiant energy from the sun but it is unusable by all organisms.

So that, it has to be converted into a usable form by photosynthetic reactions, and then transferred from one organism to another in the form of organic compounds. The series of steps through which energy is transferred from the sun to organisms (producers, consumers, decomposers) in an ecosystem called is food cahin. In a living region, there are producers, consumers and decomposers. These are like links of a chain. (*Figure-2.8*).

The absence of a link breaks the association.

Producers:

The bacteria, protists and plants that can convert light energy into chemical energy are called producers. These organisms form the first link of the food chain. For this reason, on land, the food chain generally starts with flowering plants, in aquatic places it starts with microscopic algae.

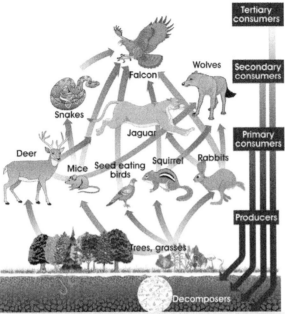

Figure–2.8.: *The food web in a terrestrial ecosystem, as in an aquatic ecosystem, starts with plants and continues with many food chains containing various animals*

Comsumers

Primary consumers: These are the animals that feed on plants, the herbivores. Examples are insects, gnawing mammals and ruminants. Mollusks and crustaceans that feed on phytoplankton in marine and freshwater are also herbivores.

Secondary and tertiary consumers: Secondary consumers are organisms that feed on herbivores, tertiary consumers are organisms that feed on secondary consumers. Animals of both groups catch their prey and have features for killing and tearing it before eating.

Decomposers:

Decomposers are mainly bacteria and fungi. These organisms have a very important role in ecosystems. For example, in forests tons of leaves are shed by trees every year. If decomposers didn't decay this layer of leaves every year it would accumulate, cover the trees and kill them.

These organisms decompose dead animals and thereby prevent grave health hazards. Nutrition starts with plants and passes to different animals. Most of the animals in a food chain feed on more than one type of food.

> All organisms need energy to live and complete their life cycle. The main source of energy is the radiant energy from the sun but it is unusable by all organisms. So that, it has to be converted into a usable form by photosynthetic reactions, and then transferred from one organism to another in the form of organic compounds. The series of steps through which energy is transferred from the sun to organisms (producers, consumers, decomposers) in an ecosystem called is food cahin.

In other words, food relations in a community or an ecosystem are not formed from a regular chain. Sometimes they contain complex interconnections of many food chains called a food web. Food webs may have short and long chains. Carnivores have a variety of food sources which causes the chain to have a complex structure and form a web. For example, falcons and eagles eat different bird species, snakes and small mammals. (*Figure-2.9*).

The consumers located at the higher levels of food chains are not always carnivores, but sometimes parasites or organisms feeding on organic wastes. The most important feature of parasitic food chains is that the organisms at the higher levels are smaller than the organisms at the lower levels. In other words, it goes from bigger organisms to smaller organisms, like dog to flea.

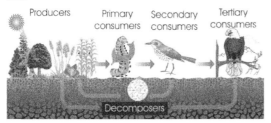

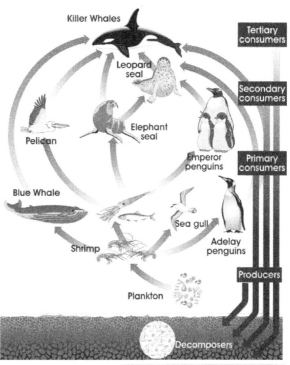

Figure-2.9.: The food web in an aquatic ecosystem. A food web, not as regular as a food chain, starts with phytoplankton and continues with various animals, sometimes interconnected with different chains, and ends in decomposers.

! Each ecosystem has a trophic structure of feeding relationships. Each level in food web is called a trophic level. The first trophic level is formed by producers, the second trophic level by primary consumers (herbivores), the third trophic level by secondary consumers (carnivores).

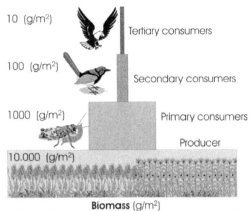

Figure–2.10.: *Biomass pyramid. Biomass decreases from producers to consumers. The organisms in the chain convert only 10% of the energy in food into biomass. Biomass decreases up to the end of the chain or pyramid.*

Ecological Pyramids

The values of some ecological factors can be shown in a pyramid for a concrete explanation. Examples are energy pyramids and biomass pyramids. Ecological pyramids are prepared on the basis of biomass, which includes the number of individuals of the community and ecosystem, and energy. Biomass of terrestrial animals is 1% of the biomass of terrestrial plants. More than 90% of this animal mass is invertebrates.

Pyramids of biomass

Biomass means "living weight". Biomass is a quantitative estimate of the total mass or amount of living material in a particular ecosystem. For example, the total weight of the roots, stems and spikes of wheat in a one hectare wheat field is called biomass. Organisms may be either plant biomass or animal biomass. (*Figure-2.10*).

Pyramid of numbers:

It shows the total number of organism at each trophic level in a given ecosystem. Let's explain this with an example. Plant-Grasshopper-Frog-Trout-Human When you look at the food chain above carefully you will see that a human is at the end.

According to this food chain, if we conclude that a human needs 300 trout per year as a food source, the trout must consume 90,000 frogs, the frogs 27,000,000 grasshoppers, and the grasshoppers 1000 tons of plants yearly. Using these values, let's form a food pyramid.(*Figure-2.11*). If the food chain above is shortened eliminating the trout, then 90,000 frogs would feed 30 people yearly. If frogs and grasshoppers were eliminated too, then 1000 tons of plants would feed 2000 people yearly.

Figure–2.11.: *As seen in the figure, the number of individuals is highest at the bottom of the pyramid and lowest at the top.*

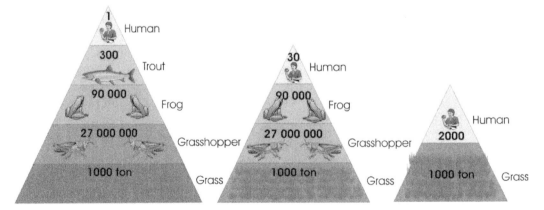

In short, the shorter the food chain the less energy lost. As can be under-stood from this data, the lowest layer of the pyramid has the greatest number of individuals. Photosynthetic organisms use only 1% of light energy in photosynthesis. Grasshop-pers convert only 10% of the ingested food into biomass. Most of it is excreted undigested or used for energy. Likewise, other organisms and humans convert 10% of the ingested food into biomass. This feature is true of all layers of all food chains. As we mentioned before, as the number of individuals in the food pyramid decreases, food and energy flow decrease accordingly. Some poisonous substances like DDT, cyanide and other chemicals cannot be excreted from the body, and their concentration increases at every level of the pyramid. (*Figure-2.12*).

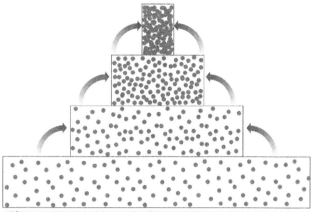

Figure–2.12.: *At every step in the food pyramid, the accumulation of chemicals increases. The organisms at the top of the pyramid are the most vulnerable to poisonous chemicals, like DDT.*

There is an inverse relationship between body size (biomass) and numbers of organisms. In other words, in a food chain, the number of large organisms is small, and the number of small organisms is large.

Pyramid of energy:

It indicates the energy content in the biomass of each trophic level. An energy pyramid is the best way to explain the flow of nutrients in an ecosystem. These pyramids demonstrate how energy is lost between layers. The total amount of energy is the greatest in the lowest layer. As you go up, energy decreases. Energy pyramids are shown as triangles because energy is lost at every level. Energy pyramids illustrate how much energy is transported to the ultimate consumers in ecosystems.

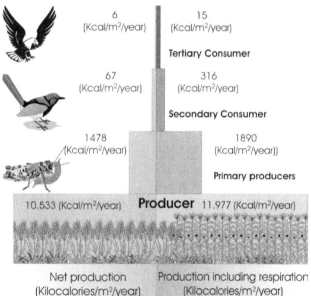

Figure–2.12.: *Food chain and energy flow. Only 10% of the energy is captured at each step from producers to consumers. Therefore, the amount of energy at the end of the chain is the lowest.*

Energy flow

The main energy source that powers natural systems is the sun. While plants utilize solar energy directly through photosynthesis, animals make use of it indirectly. Energy is present in various forms in nature, such as mechanical, chemical, electric, nuclear, heat and light energy. Living things need all of these except nuclear energy. Energy must be converted from one form to another for the continuity of life. For instance, a person walks because the chemical energy of food is converted to mechanical energy. After energy is used to perform body functions, the remaining energy is heat energy.

As can be seen in (**Figure-1.23**), the organic substances produced by green plants are called primary products. Herbivores that feed on the primary products form organic substances called secondary products. Carnivores that feed on secondary products form organic substances called tertiary products. Generally 90% of the energy is lost from one layer to the next, in accordance with the second law of thermodynamics. Only 10% of the energy is transferred to the next layer. This energy is called usable energy, and biologists refer to the "10% law". Consequently, energy flow is the greatest at the beginning of the food chain, and smallest at the end. The remaining energy is lost as heat.

Energy flow in ecosystem. The main source of energy for all organisms is solar energy. In photosynthesis, solar energy is converted to chemical energy that can be used by plants. Some of this energy is used in the metabolism of the organism and the rest is dissipated into the environment as heat. The level of energy is the highest in plants, producers, and the lowest in tertiary consumers.

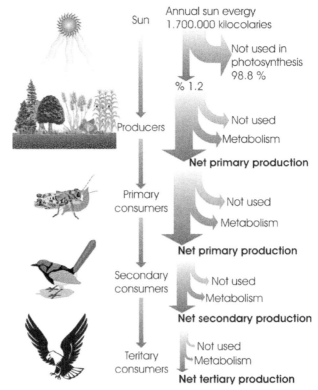

READ ME! Energy from the sun

Our sun is an ordinary star, average in size and brightness, compared to the millions of others in the universe. But when energy from the sun travels through 93 million miles of space in only eight minutes to reach us here on Earth, extraordinary things can and do happen. How does the sun make energy? The sun is a huge ball-shaped cloud of hot gases held together by gravity. It is made up mostly of hydrogen and helium. Inside the sun, hydrogen atoms moving very quickly collide with one another. Sometimes they combine to make helium atoms in a nuclear process called fusion.

During fusion, a tiny amount of mass is lost. One helium atom weighs just a little bit less than two hydrogen atoms. That little bit of mass is transformed into an enormous amount of energy, mainly infrared and visible light, which radiates in all directions through space. The sun has been emitting energy constantly for about five billion years. Astronomers estimate it will continue for another five billion. Only a small fraction of solar radiation (one part in two billion) reaches the earth. Even so, the sun is the source of almost all the energy on earth, including our food and our fuel.

Every day, Earth is bombarded by about 1022 joules (j) of solar radiation (1j = 0.239 calories). This is the energy equivalent of 100 million atomic bombs the size of the one dropped on Hiroshima. The amount of solar radiation reaching the globe ultimately limits the photosynthetic output of ecosystems, although photosynthetic productivity is also limited by water, temperature, and nutrient availability.

Much of the solar radiation that reaches the biosphere lands on bare ground and bodies of water that either absorb or reflect the incoming energy. Only a small fraction actually strikes algae, photosynthetic bacteria, and plant leaves, and only some of this is of wavelengths suitable for photosynthesis.

Of the visible light that does reach photosynthetic organisms, only about 1% to 2% is converted to chemical energy by photosynthesis, and this efficiency varies with the type of organism, light level, and .other factors. Although the fraction of the total incoming solar radiation that is ultimately trapped by photosynthesis is very small, primary producers on Earth collectively manufacture about 170-200 billion tons of organic material per year-an impressive quantity.

CHOOSE THE CORRECT ALTERNATIVE

1. Which of the following is the source of energy for nearly every organism in nearly every ecosystem?

   A. Radiant energy
   B. Geothermal energy
   C. Wind energy
   D. ATP energy
   E. None of the above

2. Which of the following organisms are secondary consumers?

   A. Plants   B. Herbivores   C. Green algae
           D. Cows   E. Lion

3. How many percent of the energy at one trophic level is passed on to the next highest trophic level?

   A. 0-5   B. 5-10   C. 15-30
           D. 30-50   E. 50-100

4. Where do plants get the carbon they use to make organic molecules?

   A. Water
   B. Carbon dioxide
   C. Glucose
   D. Oxygen
   E. Starch

5. 100,000 kcal of producer could support approximately ____ kcal of tertiary consumer.

   A. 100   B. 1000   C. 10,000
           D. 100,000   E. 1,000000

6. When you eat a banana, you are a ____ ____.

   A. primary producer
   B. tertiary consumer
   C. secondary consumer
   D. primary consumer
   E. carnivore

7. Which of the following are photosynthetic organisms?

   A. Consumers   B. Autotrophs
   C. Aeterotrophs   D. Chemotrophs
               E. Carnivores

8. Which of the following organism are the main decomposers in an ecosystem?

   A. bacteria and animals
   B. plants and animals
   C. prokaryotes and animals
   D. fungi and bacteria
   E. plants and fungi

9. Which of the following is a primary producer?

   A. Apple tree   B. Lion
   C. poison frog   D. Eagle
               E. Cat

10. Which of the following is an example of mutualism?

    A. Prey and predator
    B. Host and parasite
    C. Lichen
    D. Cat and mouse
    E. All of the above

ECOLOGY

POPULATION ECOLOGY

Chapter 3

POPULATION ECOLOGY

Population is a group of organisms of the same species that live in the same geographical area at the same time. Populations are the smallest units of communities. The word population was actually first used for people. Later it was also used for other living organisms. The individual members of the population interact among themselves and with the other living things around as well.

Populations are organized into communities A **community** is the sum of populations in a given area. A community ecologist might study how organisms interact with one another-including who eats whom and interaction between organisms and nonliving environment.

Ecosystem is the interacting system that contains a community and its nonliving physical environment.

So, an ecosystem includes not only all of the interactions among the living organisms of a community but also all of the interactions between the organisms and their physical environment.

An ecosystem ecologist for example might examine how temperature, light, precipitation and soil factors affect the organisms living in a tropical rain forest or desert. All of the ecosystems of living things on earth are organized into biosphere. The organisms of biosphere depend on one another and on other divisions of earth physical environment.

> People initially showed little interest in the study of ecology, probably because environmental problems weren't as acute and there weren't as many threatened and endagered species as there are today. Though the first ideas about ecology were recorded in 300 B.C, ecology was not accepted as a separate branch of science until the 19th century. The word "Ecology" was first used in 1867 by Ernst Haeckel.

Features of Population

There are five main features of a population.

- Population size
- Population density and carrying capacity
- Age distribution in population
- Variations seen in populations
- Dispersion in population

Population size and growth

Several factors cause changes in population size. Birth and immigrants (the organisms that move into a population) increase the size of population. Deaths and emigrants (the organisms which move out of a population.) decrease the size of population. If the number of organisms that are born plus the number of immigrants is bigger than the number of organism that die plus the number of emigrants the population will increase in size is called population growth.

Changes in the population size	=	Natality + Immigration	−	Mortality + Emigration
X		Y		Z

According to this formula;

- If Y > Z, population increases.
- If Y < Z, population decreases.
- If Y = Z, population is balanced.
- If Y decreases continuously, the population is faced with extinction. Especially if this decrease occurs as a result of high mortality rather than high emigration, the species will soon become extinct.

The world population is the total number of humans alive on the planet Earth at a given time. According to estimates published by the United States, the Earth's population reached 6.5 billion on Saturday, February 25, 2006.

The 20th century saw the biggest increase in the world's population in human history. The following table shows estimates of when each billion milestone was met:

1 billion was reached in 1802.

2 billion was reached 125 years later in 1927.

3 billion was reached 34 years later in 1961.

4 billion was reached 13 years later in 1974.

5 billion was reached 13 years later in 1987.

6 billion was reached 12 years later in 1999.

These numbers show that the world's population has tripled in 72 years, and doubled in 38 years up to the year 1999.

READ ME!

Population growth rate

Populations of organisms change over time. On a global scale, this change occurs by two factors; the number of birth and the number of deaths and numbers of immigrants and emigrants in the population.

For example in human population, the birth rate is expressed as the number of birth per 1000 people per year, and the death rate is expressed as the number of deaths per 1000 people per year.

The rate of change, or growth rate(r), birth rate (b), death rate (d)

Formula $r = b - d$

As an example, consider a population of 100,000 members in which there are 10000 births per year (or 100 births per 1000 people) and 5000 deaths per year(or 50 deaths per 1000 people)

$r = b - d$ $r = (100/1000) - (50/1000)$ $r = 0.1 - 0.05 = 0.05$

$r = 0.05$ means the population has an annual growth rate of 5%.

Doubling time

It is the amount of time it would take for the population to double in size. If the growth rate does not change doubling time can be calculated by the below formula.

$T_d = 0.7/r$ So in our example $r = 0.05$ doubling time would be

$T_d = 0.7/0.05 = 14$ years.

If we add to the equation, the number of immigrants and emigrants, the formula can be like this.

Emigrants (e)   Immigrants (i) $r = (b - d) + (i - e)$

For example, the growth rate of a population of 10000 that has 1000 births, 500 deaths, and 10 immigrants, 100 emigrants in a given year can be calculated as follows:

$r = (b - d) + (i - e)$ $r = (1000/10000 - 500/10000) + (10/10000 - 100/10000)$

$r = (0.1 - 0.05) + (0.001 - 0.01)$ $r = 0.05 - .009 = 0.041$

The positive number means population becoming bigger.

Types of Population Growth

A line graph of a population's growth through time is called a population growth curve. The population grows slowly at first but then grows more rapidly. The population size finally becomes stable at a certain level. The population size may then change slightly, but it remains around the same level.

Population growth is examined in two ways:

- **S-Shaped growth curve (Sigmoid)** In the S-shaped curve, the population increases slowly at first (construction phase), and then population enters a period of exponential growth (logarithmic growth phase).

The rate of growth gradually slows down as the population approaches the carrying capacity, forming the top of the S-shaped curve, at which point the population is balanced. (*Figure-3.1*).

Populations with an S-shaped growth curve have logistic growth. Let's examine this growth in detail. Initially the population growth rate is low because such problems as low numbers of individuals, and habitat and shelter deficiency result in less reproduction. After the construction phase, in the logarithmic growth phase, the natality rate is higher than the mortality rate. There are more reproducing organisms, and individuals have prepared shelter for their offspring. This condition continues for some time until the population growth rate begins decreasing. This is due to environmental resistance, increase in parasitic diseases and increased competition. These factors limit growth and cause recession.

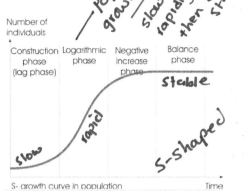

Figure–3.1.: The graph above shows an S-shaped population growth curve. Population is low in the construction phase and later increases. A short recession phase follows and then equilibrium is reached.

In sociology and biology, a population is the collection of people, or organisms of a particular species, living in a given geographic area, or space, usually by census. In biology, plant and animal populations are studied.

- **J-Shaped Growth Curve** This second type of population growth, a J-shaped curve, is seen in some insects. A short construction phase leads into a logarithmic phase, but when environmental restraints reappear, the population quickly declines. In this way there is no balance in the J-shaped curve. Other phases are similar to those of the S-shape.

These organisms may reach a maximum number of individuals in hot places, but when cold weather arrives, they enter into an extinction phase before reaching their balance. In other words, population recedes under adverse environmental conditions. (*Figure-3.2*).

For example, fly, butterfly, and mosquito populations increase to a maximum in the warm weather from April to August. With the start of winter, their populations decline.

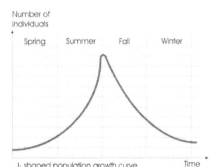

Figure–3.2.: A J-growth curve is seen mostly in the growth of invertebrates. In such populations, environmental resistance causes recession in the growth.

Population Density and Carrying Capacity

Population density is the number of individuals of a species that live in a particular area, and biomass implies the biological weight. For example, the density of mammals or trees in a region is the number of individuals or trees per hectare or km^2; for invertebrates, the number of individuals per m^2; for aquatic organisms, the number of organisms per m^3 is used to express population density.

The determination of population density is very important in ecology because the role of the population in the ecosystem depends on its density. At the same time population density implies population size.

The number of individuals per unit area or volume is related to the size of the organism, and generally small organisms have a higher population density than large organisms.

The combined limitations of the environment form a ceiling on the number of individuals that can be supported in an area. The maximum size of a population that can be supported indefinitely is called the carrying capacity of the environment.

Over time the carrying capacity of the population changes continuously. Seasons determine the increase or decrease of the carrying capacity. (*Figure-3.3*).

The maximum rate of population growth under ideal conditions is called Biotic Potential. Extreme increase in population density results in strong competition for food, shelter, and mates. Older, weaker organisms lose out to younger, stronger organisms.

At the same time rapid increase leads to an increase of predators, parasites and disease. When density is lower, food, shelter and mates are more available and for this reason population density increases. (*Figure-3.4*).

Both low and high population density are dangerous for social organisms. Cooperation and division of labor are very vital to these organisms. In honeybees, for example, the temperature of the hive is maintained by a certain density. Cooperation of these bees provides constant temperature in autumn. The hive is cooled by the beating wings of the bees and warmed by their body heat.

Agricultural pests and weeds are better controlled with biological measures than chemical measures. Biological controls decrease the population density of pest species, and so their rate of reproduction decreases. Since predators mostly hunt particular species, the population will disappear.

Figure-3.3.: *The first graph shows an ideal S-shaped growth curve. The second graph shows a sudden population increase where the population significantly exceeds the carrying capacity. As a result of environmental resistance, the population decreases. Later it stabilizes at the new carrying capacity.*

Figure-.3.4.: *Change in the number of individuals in a population over time.*
- *In regions 1 and 3 there is population growth. The number of individuals added is greater than the number of individuals leaving the population.*
- *In regions 5 and 7 the number of individuals leaving the population is higher.*
- *In regions 2, 4 the population is in equilibrium.*
- *In region 4 the population has reached the carrying capacity.*
- *In regions 2 and 6, environmental conditions providing stability are not the same because the number of individuals is different.*

Age distribution in Population

Age distribution, which affects the death rate, is a very important feature of a population. The mortality rate changes with age. Reproduction is seen only in certain age groups. In fast growing populations, there is a high ratio of young and middle-aged individuals. Balanced populations have the same ratio of all age groups. Declining populations have a higher ratio of old individuals.

Population has three ecological phases: pre-reproductive, reproductive and post-reproductive. In developed populations of humans, each of the three phases is present in the ratio of 1/3. Insects, however, have a long pre-reproductive phase and a short reproductive phase. They have no post-reproductive phase at all.

In human populations, the pre-reproductive phase is from 1-15 years of age, the reproductive phase is from ages 15-59 years, and the post-reproductive phase includes those over 60 years old. The ratio of these groups to the total population varies from country to country.

Age distribution in populations is illustrated by polygons and age diagrams. A polygon with a wide base implies a high ratio of young people in that population. (*Figure-3.4*).

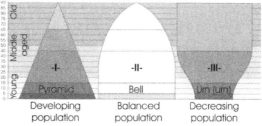

Figure–3.4.: *Age distribution in populations is shown with diagrams. The base of the pyramid shows children (infertile, pre-reproductive), the middle part shows adults (reproductive), and the top part shows elderly (infertile, post-reproductive). Pyramid shows a growing population, Bell shape shows a balanced population, Urn shape shows a decreasing population.*

Every species has a certain lifespan. Lifespan varies from one species to another. Those that adapt to their habitat complete their maximum lifespan but, due to environmental factors, some die without completing their lifespan. This is illustrated in the graph below.

Generally, survivorship curves are used to show lifespan of species. Survivorship is the proportion of individuals in a population that survive to a particular age.

When the graph is examined, it is clear that Drosophilae (*Type-I*) are well-adapted to their habitat and many reach their maximum lifespan. Humans (*Type-I*) are also well-adapted.

The mortality ratio of Hydra (*Type-II*) is stable throughout their life. Lobsters (*Type-III*) encounter high environmental resistance especially in the larval stage, when they are eaten by many other organisms, and early death is observed. (*Figure-3.5*).

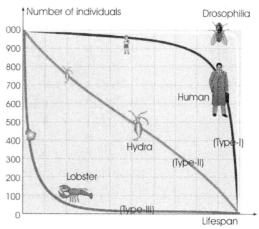

Figure–3.5.: *Survivorship curves show lifespan and age distribution of some species.*

Figure-.3.6.: *Prey-predator relations. The size of some populations is limited by another population in the same habitat. In this relation, as predators increase, prey decreases. Again as predators decrease, prey increases.*

Fluctuations in Population

Over time populations may encounter sudden increases or decreases in growth. Such changes are called fluctuations. In some populations these fluctuations don't happen and the number of individuals is stable when the population reaches its balance.

For example, the human population is balanced as long as there are no wars, earthquakes, floods or other such calamities. The size of some populations is limited by another population in the same habitat, a predator-prey relationship for example

For example, a rabbit population is affected by lynx, a type of big cat. Lynx prey and feed on the rabbits. As a result, as the lynx population increases, the rabbit population decreases. (*Figure-3.6*).

This situation continues for a certain time. Eventually the lynx can't find enough food and their population declines seriously due to a high death rate, migration and competition. The decrease of the lynx population results in the increase of the rabbit population. In conclusion, as predators decrease, prey increases; as prey decreases, predators increase.

Competition

Competition is the struggle between two organisms or two populations to obtain the same resources. Both groups are harmed, but the degree of damage is different. In other words, one side can destroy the other side. Some resources for which organisms compete are food, light, shelter and hiding places. (*Figure-3.7*).

Ecological competition may be between the members of the same species or different species. For example, the struggle of two dogs for a bone, and the struggle of hens for the same food are examples of intraspecific competition.

The struggle of a cat and a dog for a piece of meat is an example of interspecific competition. Ecological competition has an important role in the development of populations.

Scientific research focuses mainly on interspecific competition because intraspecific competition is mostly due to factors related to population density.

Population distribution

The individuals comprising a population often exhibit characteristic patterns of distribution, or spacing. Distrubution means migration of the population. The most significant reason for migration is adverse changes in habitat, but there are also dispersions related to mating and reproduction. For example, the seasonal migration of birds, and the migration of fish from freshwater to the sea are notable. Migration is important because it causes the population to decrease. (*Figure-3.8*).

Figure–3.8.: *Distribution types in populations a) Clumped distribution b) Uniform dispersion c) Random distribution*

Populations exhibit different patterns of distribution, namely, uniform, random and clumped.

- **Uniform dispersion:** occurs when individuals are more evenly spaced. This is seen very rarely in nature.

- **Clumped dispersion:** occurs when individuals are concentrated in specific portions of the habitat. This is the most common spacing seen in nature. This may be because of patchy distribution of resources. In every clump there may be different numbers of individuals. Fishes; humans and birds are spaced like this (*Figure-3.9*).

- **Random dispersion:** occurs when individuals in a population are spaced in an unpredictable or random way that is unrelated to the presence of others.

Figure–3.9.: *Distribution types in populations. Clumped distribution of fifhes.*

 The UN estimated in 2000 that the world's population was then growing at the rate of 1.4 percent (or 91 million people) per year. This represents a decrease in the growth rate from its level in 1990, mostly due to decreasing birth rates.

The first five years of the 21st century saw something of a decline in the overall population growth, with the world's population increasing at a rate of about 76 million people per year as of 2005.

> In unstable or unpredictable environments r-selection predominates, as the ability to reproduce quickly is crucial, and there is little advantage in adaptations that permit successful competition with other organisms (since the environment is likely to change again).
>
> In stable or predictable environments K-selection predominates, as the ability to compete successfully for limited resources is crucial, and populations of K-selected organisms are typically very constant and close to the maximum that the environment can bear.

Reproductive tactics in populations

Each species has its own life history strategy-its own reproductive characteristics, body size, habitat requirements, migration patterns, and so on-designed around this energy compromise. Although many different life histories exist, living organisms often fall into two main groups with respect to their life history strategies as *r-selected* species and **K-selected** species.

Features	r-selected Opportunistic Populations	K-selected Equilibrial Populations
Maturation time	Short	Long
Lifespan	Short	Long
Death rate	Often high	Usually low
Number of offspring produced per reproductive period	Many	Few
Number of reproductions per lifetime	Usually one	Often several
Timing of first reproduction	Early in life	Late in life
Size of offspring eggs	Small	Large
Body size	Small	Large
Habitat	Variable environments	Stable environments
Parental care	None	Often extensive
Example	Insects	Mammals

READ ME! Shaping Populations

People residing in a certain area, pine trees in a forest, and frogs in a stream are examples of populations. Human interactions with the surrounding organisms are as important as the chemical and physical properties of the earth. Because no organism can survive by itself, it must be in association with other organisms to obtain food and other things it needs for life. Two hypotheses are postulated related to the natural factors shaping populations.

Malthus Hypothesis: Population growth is affected by external forces like war, famine, disease and natural calamities. The Malthus model states that while population growth increases geometrically, food increases arithmetically. As can be seen from the graph opposite, the growth of population is faster than the increase of the food supply. As time passes, the discrepancy gets bigger. At the present time, however developments in food engineering, agriculture and veterinary medicine have increased the quality and production of crops and livestock.

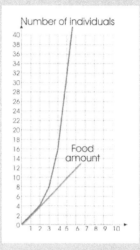

Food production can be increased especially by unraveling the mystery of photosynthesis, widening and increasing the use of biotechnology, and by spending less money on weapons and using the savings to benefit people. Famine and malnutrition throughout the world may thus be prevented.

Malthus hypothesis. According to Malthus food increases arithmetically (1, 2, 3,...) while population increases geometrically (2, 4, 8, 16..).

Wayn Edward Hypothesis: According to this hypothesis, every population regulates itself according to the food available so that there is never famine. In this way population regulation is achieved as a result of internal forces due to social behaviors like birth control.

CHOOSE THE CORRECT ALTERNATIVE

1. Which of the following is a population?
   A. all the insects that live in your home
   B. all the plants that live near each other in a forest
   C. all the types of microorganisms that live in or on your body
   D. all the mammals on Earth
   E. all of the lions that live in a particular forest

2. Which of the following pattern of dispersion is seen in wild populations mostly?
   A. clumped
   B. random
   C. equilibrial
   D. uniform
   E. none of the above

3. Which of these organisms has a survivorship curve similar to that of butterfly?
   A. cats
   B. elephants
   C. mosquito
   D. lions
   E. humans

4. Which of the following statements describes population density?
   A. 689 people
   B. 1881 organisms (lions, giraffes, hyenas and eagles) in a forest
   C. the number of sea stars in ocean
   D. 589 oak trees in a 65-square-kilometer forest
   E. the total dry weight of green algae in a lake

5. Ardahan city has a population of 38.000 individuals. They have 830 births, 450 deaths, and 10 immigrants, 100 emigrants in year. How many people were added to Ardahan population in one year?
   A. 380 B. 420 C. 470 D. 530 E. 810

6. The population has an annual growth rate of 7%. Which of the following is the doubling time of this population?
   A. 7 years B. 10 years C. 14 years
   D. 17 years E. 20 years

7. Which of the following could be a density-dependent factor that might limit population growth?
   A. Fire
   B. Storms
   C. Temperature changes
   D. Food supply
   E. Volcano

8. In the models that describe population growth, r stands for...
   A. Death rate.
   B. Doubling time.
   C. Total number of individuals in the population.
   D. Growth rate.
   E. Carrying capacity.

9. A broad-based, pyramid-shaped age-structure diagram is characteristic of a population that is...
   A. Growing rapidly.
   B. At carrying capacity.
   C. Stable.
   D. Limited by density-dependent factors.
   E. Shrinking.

10. Which of the following organism has r-selected reproductive strategy?
    A. Cat B. Lion. C. Elephant.
    D. Mosquito E. Wolf.

ECOLOGY

44

BEHAVIORAL ECOLOGY

Chapter 4

BEHAVIORAL ECOLOGY

All the actions of an organism make up its behavior. Behavior has a genetic basis and it can be observed and described. If you have ever observed a lion hunting a deer or a bird building a nest, you have observed animal behavior. **Ethology**- is the sub-branch of biology - study of animal behavior based on the systematic observation, recording, and analysis of how animals function, with special attention to physiological and ecological and aspects. Laboratory or field experiments designed to test a proposed explanation must be rigorous, repeatable, and show the role of natural selection.

Animal behavior that involves interaction between members of a population is called social behavior. We can define social behavior as the interaction of two or more animals, usually of the same species. Many animals benefit from living in groups. Two bears fighting each other is an example of social behavior. Some species that engage in social behavior form societies. (*Picture-4.1*).

What is behavior and how it works?

Picture-4.1: *Two bears fighting each other is an example of social behavior.*

Behavior is a series of activities that occur in response to a stimulus. Many animals have inherited cyclic behavior patterns that vary over the course of day, month, or year. It is thought that these patterns are regulated by a kind of internal "biological clock".

Examples of animal behavior studies could include research to determine how animals find and defend resources, communicate, avoid predators, choose mates, reproduce and care for their young. Animal behavior has both genetic and environmental components. Some behaviors develop similarly, regardless of the environment, while others develop differently in different environments.

Behavior results from both genes and environmental factors. In other word behavior capacity of an animal is inherited and can be modified by environmental factors. Some behavior patterns are innate some are learned.

Starting from simple bacteria to the giant blue whale, many organisms live in our unique biosphere and they have direct or indirect interaction between abiotic components of the ecosystem and among each other. All of these organisms are being effected by others and nonliving things, and they directly or indirectly effect others. New scientific studies showed that even the simplest cell (e.g. bacteria) show quite complex behaviors beyond simplicity of their structure.

What kind of interaction do these organisms have?

For example:

- What happens when regular intensity of the light received by a plant changes?

- How does it react color change of light?

- In what way do animals react when they see a male or female from their species?

- Why do plants grow toward certain direction?

Picture-4.2: *Tropism is directional movement, in phototropism plants move toward light source.*

> All the actions of an organism make up its behavior. Behavior has a genetic basis and it can be observed and described. If you have ever observed a lion hunting a deer or a bird building a nest, you have observed animal behavior. Ethology- is the sub-branch of biology - study of animal behavior based on the systematic observation, recording, and analysis of how animals function, with special attention to physiological and ecological and aspects.

Reaction or responses of the organisms against stimuli coming from their internal and external environment (from living or nonliving) are termed as behavior. Ethology is the branch of biology which studies behavior of organisms. Living organisms react environmental stimuli in different ways. For example a dog reacts a sound by its tail and ears, while a bird chirps (sound).

Basically the source of behavior is hereditary as potential. In some organisms this genetic program is rigid for some kind of behaviors, while in most organisms it can be improved and even changed by experience. Learning is the improvement of the behavior depending on experience in most organisms (in human experiences can be shared by others). Behaviors are usually based on protection, nutrition and reproduction. Unicellular organisms also have complex behaviors. Their behaviors are mostly related with light, gravity, food, temperature and predators. Some times all of these factors may play role and they initiate spore formation (when they become not relevant). Unicellular organisms move by cilia, flagella or Cytoplasmic projections. Some of them can not move actively, their movement is based on movement of the surrounding in which they live. Plants do not have sense organs or a communication structure. But they also have some behavior which depends on gravity, light, temperature, length of day time, humidity, water and even wind. Different stimuli result different behaviors in different plants. For example some plants move toward light, while some others move away. Most Plant behaviors are termed as **tropism** that has a direction. Depending on the factor it is named as Phototropism (light), geotropism (gravity) etc. Tropism can be positive "**toward**" or negative "**away**". Nastic movements are the second types of behaviors which have no direction. Some physical stimuli (pressure, light, temperature et,) cause some responses that has no any direction. For example when you touch mimosa plant, it closes its leaves, some plants close their flowers during the day and open them evening. Wind also makes some plants close or roll their leaves to decrease transpiration and physical damage.

Types of behaviors

Innate behaviors

The term instinct refers to Innate (inborn behavior) behaviors that are genetically programmed behaviors have not been learned. It occurs automatically.

Innate behavior can be seen in plants such as simple growth movements toward or away from stimuli (tropism), action of Venus's- fly trap, and sensitive plant "mimosa".

Examples of innate behavior in animals include hibernation in black and brown bears, nest building in birds, web weaving in spiders, and mating. Innate behavior in animals can be observed.

Picture-4.3: *innate behavior in animals*

For example newly hatched, still-blind birds beg for food by raising their heads, opening mouths and cheeping loudly when a parent lands on the side of the nest. Gull chicks peck at red spot on parent's bill to get food; chicks will also peck at artificial bills. (*Picture-4.3*)

Learned behavior

The other kind of behavior is learned behavior that results from experience and involves some choice of responses to give stimulus. Learning is an experience-based modification of behavior and can be stored in the brain as a memory and can be recalled.

Habituation

Learning has several form the simplest type of learning is called habituation. In this type of learning an organism learns no to respond to repeated "**irrelevant**" stimuli. When a stimulus occurs many times without any consequence the animal usually decreases its response. (*Picture-4.4*)

For example birds nesting near highways learn not to respond to traffic noise and young chicks learn not to run from innocuous, common stimuli, such as bowling leaves and other non-predatory birds.

Human can learn to perform many complex activities with the little or no thought such an act is called habit. Such as dressing, writing, talking and dancing. These kinds of activities are first learned and then, by frequent repetition, become automatic.

Picture-4.4: *In psychology, habituation is an example of non-associative learning in which there is a progressive diminution of behavioral response probability with repetition of a stimulus.*

Figure-4.1: *Pavlov's experiment showed that dogs can be classically conditioned to salivate at the sound of bell.*

Figure-4.2: *Skinner box, in which a rat learns to press a lever in order to obtain food.*

Figure-4.3: *Trainers teach animals to perform tricks by rewarding them for desired behavior and punishing them for undesired behavior. Elephants can learn playing football even drawing a picture.*

Conditioning

Conditioning is a type of learning that an animal learns to respond in a customary way to a new stimulus. Many animals can learn to associate one stimulus with another. Associative learning is the ability of many animals to learn to associate one stimulus with another.

Classical conditioning

Classical conditioning is a type of associative learning. It was first studied by the Russian physiologist.(*Picture-4.1*)

Ivan Pavlov (1849-1936). Pavlov's dog is a good example. Ivan Pavlov exposed dogs to a bell ringing and at the same time sprayed their mouths with powdered meat, causing them to salivate. Soon, the dogs would salivate after hearing the bell but not getting any powdered meat.

House pet often become conditioned in a similar manner. Pet owner know that the sound of the refrigerator door can cause their pets to be quite attentive and to salivate.

Operant conditioning

Operant conditioning is called trial-and-error learning an animal learns to associate one of its own behaviors with a reward or a punishment.

Trial and error learning can refine natural behaviors, such as predatory skills. For example a bear might learn that splashing about in a stream does not yield a salmon dinner. Staying still and quite in one place is much more effective. Animal trainers use operant conditioning.

Trainers teach animals to perform tricks by rewarding them for desired behavior and punishing them for undesired behavior. To the animals the absence of an expected reward is the same as a punishment.

Pigeons can learn to play table tennis in this way. Ethnologist **B.F. Skinner** did extensive research with animals, notably rats and pigeons to demonstrate Operant conditioning. He invented the famous Skinner box, in which a rat learns to press a lever in order to obtain food. (*Figure-4.2-3*)

Figure-4.4: *Konrad Z. Lorenz being followed by his imprinted geese*

Imprinting

Imprinting is a term which describes a learning process during sensitive periods of life by which organisms learn their preferences toward and a particular object or class of objects. It typically involves an animal and person learning the characteristics of some stimulus, which is therefore said to be "imprinted" onto the subject. This type of learning that occurs in young animals of certain species was discovered by in 1935 by **Kondrad Lorenz** an Austrian biolog. Lorenz observed that the young of birds such as geese and chickens spontaneously followed their mothers from almost the first day after they were hatched, and he discovered that this response could be imitated by an arbitrary stimulus if the eggs were incubated artificially and the stimulus was presented during a critical period (a less temporally constrained period is called a sensitive period) that continued for a few days after hatching.

Insight

Insight is the ability to find a solution an unfamiliar problem without a period of trial and error. In other worrd insight learinig is the ability of an animal to perform a correct or appropriate behavior on the first attempt in a situation with which it has had no prior experience. For example, in one experiment hungry chimpanzees were released into a room with boxes and bananas hanging from the ceiling out of reach. They find that it is not possible to reach the bananas by jumping and chimpanzees reasoned that stacking boxes would enable them to reach the bananas. They solved the unfamiliar problem by insight learning.

Figure: *No comment*

> A society is a self-reproducing grouping of individuals occupying a particular territory, which may have its own distinctive culture and institutions. As culture is generally considered unique to humans, the terms "society" and "human society" have the same meaning. "Society," may refer to a particular people, such as the Nuer, to a nation state, such as Austria, or to a broader cultural group, such as Western society.

Society and its features

A society is an actively cooperating group of individuals belonging to the same species. A hive of bees, a flock of birds, a pack of wolves are examples of societies. Some societies are loosely organized, whereas others have a complex structure. Characteristics of a well-organized society include cooperation and division of labor among animals of different sexes, age groups, or castes.

Communication in society

Social interactions depend on communication, the transmission of information between individuals. Animals communicate with one another by means of vvisual, auditory (sound), tactile (touch), chemical or electrical signals, or by chemical signals called pheromones.

Transmitting and receiving these signals generates an external or internal response in the organism. Pheromones, chemical signals used for communication, are especially common in mammals and insects. Bees use pheromones to determine social rank and to initiate reproduction.

The context of the signal is important. For example, male honeybees will respond to the pheromones of the queen while they are outside the hive (where they can mate) but are unaffected by the pheromones while inside the hive. Ants also use pheromones to mark a trails to food sources.

> The first time you try out a camera you may not need someone to explain how to work it. Instead you may use what you already know about other cameras to figure out how the new one works.
>
> When you solve a problem or learn how to do something new by applying what you already know, without a period of trial-and-error-, you are using insight learning.

READ ME! Communication in honeybees

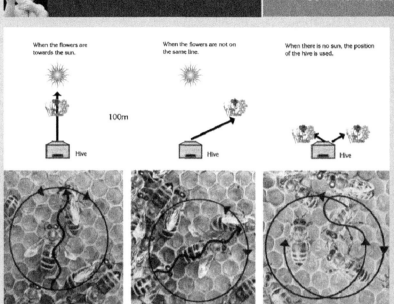

One of the most Complex communication systems certainly among invertebrates is that of social or hive, bees. Pheromones produced by a hive's queen and her daughters the workers; maintain the social order of honeybee colonies.

How do bees communicate?

The problem was first studied in the 1940s by Karl von Frisch, who carefully watched individual European honeybees (Apis mellifera carnica) as they returned to special observation hives.

Round dance: The round dance indicates that food is near but may provide no information on directionality or specific distance. A returning bee would quickly become the center of attention by other bees, called followers. The returning bee would go through a repetitive behavior that von Frisch called a dance. If the food source was close to the hive (less than 50m away), the returning bee moved in tight circles while waggling its abdomen from side to side This dance was usually accompanied by the bee's regurgitating some of the acquired nectar, this behavior, which von Frisch called the round dance, had the effect of exciting the follower bees and motivating them to leave the hive and search for food that was nearby.

Waggle dance: However, bees often forage at great distances from the hive, sometimes in excess of 5 km. In such cases, the round dance is insufficient, lacking both directionality and distance cues necessary for the followers to locate the food source efficiently.

A worker returning from a longer distance does a "waggle dance" a halt circle swing in one direction, followed by a straight run and then a "know" the type of food. The waggle dance is performed when food is distant. This dance pattern resembles a figure eight, with a straight run between two semicircular movements. According to von Frisch's hypothesis, the waggle dance indicates both distance and direction. Distance is indicated by the duration of each waggle run or dance and the number of abdominal waggles performed per waggle run. Direction is indicated by the angle in relation to the vertical surface of the hive) of the straight run that forms part of the dance itself. (1) For instance, if the straight run is directly upward, this signals that food is in the same direction as the sun. (2) If the straight run is directly downward; the food is in the direction opposite the sun. (3) If the angle is 30" to the right of vertical, the food is 30" to the right of the horizontal direction of the sun, and so forth. Odor cues (pheromones) and sound may also convey information about the location and type of food.

Territoriality in a society

Territoriality is a social behavior that involves defense of a limited area, often against other members of the population. Territoriality is most commonly seen during a population's breeding season. For instance, a pair of robins will drive all other robins from the area near the nest they have built. Territorial behavior is observed in many groups of animals.

Hierarchy in society

A hierarchy, or order of dominance, is established. In populations that have such a social hierarchy, each member has a position based on the amount of aggressiveness the member exhibits. Social rank may change because of death, fighting, and, occasionally, alliances between members. The leader, or dominant member of the group, is usually also the male that fathers the most offspring.

Parental care in society

Many organisms care for their young. Care of the young is an important part of successful reproduction in many species. The benefit of parental care is the increased likelihood that the offspring will survive.

Females are more likely than males to brood eggs and young and usually the females invest more in parental care.

READ ME!

Rythmic behavior

Most organisms show periodic behavior that seems to be related to periodic changes in the environment. A behavior that occurs periodically is called a rhythmic behavior. Rhythmic behaviors often are related to length of daylight versus darkness and to the change of seasons. The rhythmic behavior of organisms seems to be regulated by an internal biological clock. This biological clock is probably a series of chemical responses to environmental stimuli. The flowering of plants is an example of a periodic behavior regulated by a biological clock.

Short-term rhythmic behaviors

Circadian rhythms: Rhythms that show a cycle of about 24 hours are called circadian rhythms. Circadian rhythms mark the active and rest phases of the behavior of most animals. The behavior of

many animals appears to be organized around circadian rhythms. Diurnal animals, like honeybees and pigeons, are most active during the day. Nocturnal animals, like Flying squirrels and bats are most active during the hours of darkness.

Lunar (moon) cycle: Some biological rhythms of animals reflect the lunar (moon) cycle. The most striking rhythms are those in marine organisms that are tuned to the changes in the tides due to the phases of the moon. Such as reproduction of most cnidarians and sponges

Long-term rhythmic behaviors

Hibernation: An example of a long-term rhythmic behavior is hibernation. Hibernation is a period of dormancy accompanied by a low metabolic rate and a much lowered body temperature. Animals that hibernate during the coldest part of the year. Both cold-blooded and warm-blooded animals hibernate. Hibernation appears to be an adaptation to surviving harsh environmental conditions, such as cold temperatures and the scarcity of food.

Migration: Migration is another example of rhythmic behavior. Migration is the movement of a population to new feeding lands or to breeding lands, with the population later returning to the original area. As with hibernation, migratory behavior seems to be triggered by seasonal changes. Thus many birds in the Northern Hemisphere fly to south in the fall. In the spring the birds return north, where they breed and raise their young. Migration also occurs in some species of fish, mammals, and insects. It seems to be an adaptation that enhances reproductive success and the survival of the young.

Estivation: Another example of rhythmic behavior in response to harsh conditions is estivation. Estivation is a period of dormancy, like hibernation, during which metabolic rate and body temperature are lowered. Unlike hibernation, which occurs partly in response to cold conditions, estivation occurs in response to hot, arid conditions. Several desert mammals, such as ground squirrels and pocket mice, estivate during the hottest, driest season. Such behavior helps the organisms to avoid overheating and prevents excessive loss of water.

CHOOSE THE CORRECT ALTERNATIVE

1. Pet owner know that the sound of the refrigerator door can cause their pets to be quite attentive and to salivate. They demonstrated a pattern of learning called _.

   A. imprinting
   B. Insight
   C. Classical conditioning
   D. operant conditioning
   E. Trial-and-error learning

2. Which of the following is a chemical messenger which is used to communicate by ants?

   A. ATP   B. NADH   C. DNA
   D. Pheromones   E. Hormones

3. Animals that are active at night are described as _.

   A. Lunar   B. Diurnal   C. Circadian
   D. Night-blind   E. Nocturnal

4. Every night Kemal turns on the light in the laboratory and then feeds the piranhas in the aquarium. After a couple of weeks of this routine, Kemal noticed that the piranhas came to the surface to feed as soon as the lights were turned on. Which of the following type of behavior can explain the behavior of piranhas?

   A. Innate behavior
   B. Imprinting
   C. Trial-and-error learning
   D. Operant conditioning
   E. Classical conditioning

5. Which of the following is a type of behavior that is genetically determined and that cannot be modified?

   A. Innate behavior
   B. Habituation
   C. Trial-and-error learning
   D. Operant conditioning
   E. Classical conditioning

6. Birds nesting near highways learn not to respond to traffic noise and young chicks learn not to run from innocuous, common stimuli, such as bowling leaves and other non-predatory birds. The lack of response is an example of _.

   A. Operant conditioning
   B. imprinting
   C. altruism
   D. habituation
   E. trial-and-error learning

7. Which of the following a learning process that can occur only during a limited period of the individual's development?

   A. Instinct   B. Conditioning
   C. Imprinting   D. Habituation
   E. None of the above

8. Which of the following bee dances shows the information that a food source is nearby?

   A. Halay   B. Tango   C. Waltz
   D. Round dance   E. Waggle dance

9. Lorenz observed that the young of birds such as geese and chickens spontaneously followed their mothers from almost the first day after they were hatched. This is an example of _.

   A. habituation
   B. operant conditioning
   C. imprinting
   D. Trial-and-error learning
   E. Classical conditioning

10. ____ is the ability to find a solution an unfamiliar problem without a period of trial and error.

    A. Insight
    B. Classical conditioning
    C. Operant conditioning
    D. Imprinting
    E. Trial-and-error learning

COMMUNITY AND ECOSYSTEM

Chapter 5

COMMUNITY

Community is the most important social unit of ecology. A community consists of all of the different species that live and interact together within an area. A community may only consist of animal and plant populations or it may have other groups of organisms (*Figure-5.1*). A community may contain other communities as well. For example, a forest community has different species of organisms. The microorganisms inside the body of an organism constitute a community as well. Communities may also be called life associations or a group of species. The type and size of the community depend on the organisms in the community and the effects of environmental factors such as temperature, rainfall, moisture and food. Populations under the effect of these factors live in harmony. For this reason, from the equator to the poles, from the prairies to the hills and mountains there are different-sized communities.

Communities make up the living portion of the ecosystem. Therefore, the study of ecosystems begins with communities. An ecotone is a zone where two ecosystems overlap. The type and width of this region are very variable. In big communities it may extend for kilometers, in small communities it may be just a few meters. Because ecotones contain individuals of both species, they have a higher variation of species than the neighboring communities. Lakeshores, stream banks, ocean beaches, the entrances of caves, and forest meadows are examples of some of our favorite ecotones.

> A biome is a large, relatively distinct ecosystem characterized by a similar climate, soil, plants, and animals, regardless of where it occurs.

Generally, from an energy-flow perspective, big communities are self-sufficient but small communities are dependent on other communities. Ecological task distribution among the species in a community increases the dynamism of the community. These species generally are dominant species and mostly are composed of plants. In aquatic communities the determination of the dominant species is difficult.

Succession

Succession is community change over time. In other words, the process of community development over time, which involves species in one stage being replaced by different species, is called succession. In succession every species prepares the habitat for another species. Because changes are observed clearly in vegetation, it is perceived as a process of plants. Ecologists recognize two types of succession.

Primary succession: which occurs in areas where no community existed before. For instance, primary succession would take place on new volcanic islands, deltas, dunes, bare rocks, and in lakes, to mention a few.

Secondary succession: which occurs in disturbed habitats where some soil and, perhaps, some organisms still remain after the disturbance. Secondary succession occurs after fires, floods, drought, and some human practices (slash and burn clearing of forests, construction projects). It also occurs on abandoned farmlands, in overgrazed areas, and in forests cleared for lumber. In natural areas the order of formation in primary succession is: lichens–mosses–grasses–shrubs–trees (*Figure-5.2*).

Lichen Phase: Places like sandy, bare rock and clay, where there is no other life, are first inhabited by lichens. Lichens secrete acids that help to break the rock apart, which is how soil starts to form. Lichens also add valuable organic matter to the young soil. Lichens, though they are very resistant to extreme physical conditions, can't compete with other organisms and, once other organisms start growing, their number decreases.

Figure–5.1.: *The transition from aquatic succession to terrestrial succession is a good example of secondary succession. a) Organic substances in a lake start filling up the lake basin. b) As the lake fills, the soil around the lake slides into the lake as well. c) The accumulation of organic substances in the lake continues until only a small pond remains at the center. d–e) The lake fills completely and terrestrial succession passes to the bush and shrub phase.*

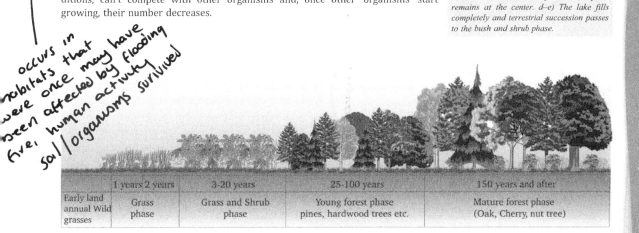

	1 years	2 years	3-20 years	25-100 years	150 years and after
Early land annual Wild grasses	Grass phase		Grass and Shrub phase	Young forest phase pines, hardwood trees etc.	Mature forest phase (Oak, Cherry, nut tree)

Sample Succession

Although succession is generally associated with plants, it is also seen in animals. Observation of succession in animals is possible in the laboratory. Though succession takes a long time in nature, it can be quickly observed in the laboratory.

A protozoa culture prepared under laboratory conditions contains many bacteria. After this the succession of protozoa increases. Here, first flagellates are seen and then paramecia.

If you examine the graph below, you will see that in the first days the dominant species is green algae. Then Colpoda that lives with algae between days 10- 20; no dominant species is observed in days 20-40; at the end of the 40th day hypotonicha becomes dominant.

After 60 days the number of hypotonicha decreases very rapidly and the number of Vorticella increases.

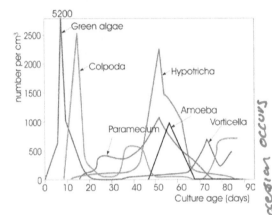

As can be understood from this experiment, food, competition, environmental pollution, temperature and other environmental factors cause succession among the species in the culture community.

Discussion:

What is the function of the plants in the culture medium?

How can you explain the change in the number of protozoa over time?

Moss Phase: The moss phase starts after the lichen phase. The most important activity of these organisms is to add moisture to the soil, after which some invertebrates move in, followed by insectivore mammals. In other words, fauna forms parallel to flora. With the development of mosses and the addition of dead organisms, soil formation speeds up and humus quality increases. In this way mosses prepare the medium for another organism.

Grass Phase: Annual grasses begin to grow in competition with the mosses. In time the number of insects increases both in quantity and variety. Reptiles, frogs, birds and mammals settle and increase in number.

Shrub Phase: The conditions created by the grasses make way for the growth and development of shrubs. These are generally small plants like berries and drupes. Another important step in this phase is the transportation and deposition of tree seeds by birds.

Tree Phase: Trees start to grow during the shrub phase. Over time, the trees grow and form a forest canopy. Shrubs may continue to grow under the canopy, but most diminish over time. In the open areas, mosses are still present. Ferns multiply in wetlands. Barring extraordinary occurrences, permanent communities of fauna and flora form. This is called climax. The climax community continues until there is some change in climate or environment, at which point it disappears. Substantial changes in the climax community, as a result of volcanic eruptions or floods, are followed by secondary succession. Secondary succession occurs for the following reasons.

- Succession begins with changing environmental conditions, deteriorating living conditions, and weakening of the competitiveness of species.
- Existing species prevent the settlement of new species.
- It is observed that animal species are especially effective on some plant species. The effect of rabbits on grass and the effect of insects on grassland can be given as examples.
- Physically, freeze, fires, storms, drought, volcanic activities, earthquakes and the effects of humans can destroy communities.

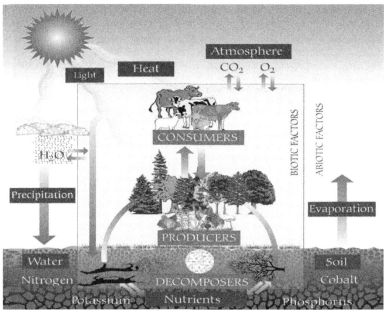 

ECOSYSTEM

Recall that, Ecosystem is the interacting system that contains a community and its nonliving physical environment. So, an ecosystem includes not only all of the interactions among the living organisms of a community but also all of the interactions between the organisms and their physical environment. An ecosystem ecologist for example might examine how temperature, light, precipitation and soil factors affect the organisms living in a tropical rain forest or desert. All of the communities of living things on earth are organized into biosphere. The organisms of biosphere depend on one another and on other divisions of earth physical environment. Some examples of ecosystem are mountain Everest, Black sea.

What is a biome?

The biosphere can be divided into regions called **biomes.** A biome is a large region that has a distinct combination of plants and animals. Climate is a factor in determining the type of biome that occurs. A terrestrial biome is usually identified by the types of plants that make up a climax community within it. The dominant types of plants are called the **climax vegetation.** However, a biome includes all stages of succession leading up to the climax community. In the deciduous forest biome, for example, deciduous trees are the climax vegetation. Ecologists have identified several biomes in the world. Ecotones are transition zones where ecosystems meet and intergrade. All aquatic and terrestrial biomes interact with one another, such that there is no isolation.

 Species richness on islands depends on island size and distance from the mainland. Because of their size and isolation, islands provide great opportunities for studying some of the biogeographic factors that affect the species diversity of communities. Imagine a newly formed island some distance from the mainland.

Robert MacArthur and E. O. Wilson developed a hypothesis of island biogeography to identify the determinants of species diversity on an island.

> An ecosystem consists of all the organisms living in a community as well as all the abiotic factors with which they interact. The dynamics of an ecosystem involve two processes: energy flow and chemical cycling. Ecosystem ecologists view ecosystems as energy machines and matter processors. We can follow the transformation of energy by grouping the species in a community into trophic levels of feeding relationships.

Biomes and climate

The main factor that determines the kind of biome in a certain area is climate. Recall that climate is determined mainly by temperature and precipitation. Average temperature decreases from the equator to the poles.

A decrease also can be seen from sea level to the mountains. Features such as mountain ranges and the nearness of large bodies of water affect precipitation. Average monthly temperature and precipitation can be plotted on a graph called a climatogram. Biomes can be identified by their dominant animal populations.

The same kind of biome is found at the same latitude, or distance from the equator, in different parts of the world. For example, there are grassland biomes in South America, Africa, and Australia. The specific plants and animals found in these different grassland biomes are not identical, however. Populations in different biomes have similar characteristics and are sometimes related, but they are different species. Plants and animals from the same type of biome resemble each other because they are adapted to nearly identical physical and climatic conditions.

Figure–5.2.: *Climate changes with latitude and topography. Different latitudes are shown on the map.*

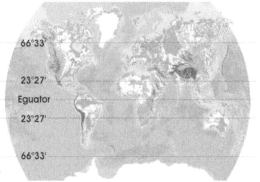

Terrestrial Ecosystems (Biomes)

All the ecosystems formed by the organisms on earth altogether are called the ecosphere. All the communities on earth are together called the biosphere. The ecosphere includes the interactions of the biosphere, atmosphere, hydrosphere and lithosphere. The populations in the ecosphere are distributed over different areas. The distribution of populations on earth is affected mainly by climate. Climatic conditions are affected by latitude and topography. Between certain degrees of latitude, major climates are seen.

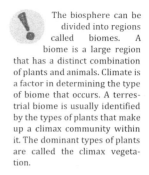

The biosphere can be divided into regions called biomes. A biome is a large region that has a distinct combination of plants and animals. Climate is a factor in determining the type of biome that occurs. A terrestrial biome is usually identified by the types of plants that make up a climax community within it. The dominant types of plants are called the climax vegetation.

These areas and climates are listed below (*Figure-5.2*).

Latitude (º)	Climate
0-23º27'	Hot (tropical) climate
23º27' – 66º33'	Temperate climate
66º33' – 90º	Cold (polar) climate

Each of these zones has a different climate. Populations of various sizes live in different climates. A large, relatively distinct terrestrial region characterized by similar climate, soil, plants and animals is called a biome. Biomes are not separated by specific boundaries and may overlap in some regions. Biomes are the biggest units of ecological systems. Usually biomes are named for their dominant plant species. The biomes are tundra, taiga (evergreen forest), deciduous forest, grassland, shrubs, tropical shrubs, savannas, tropical rainforests, semi-deserts and deserts (*Figure-5.3*).

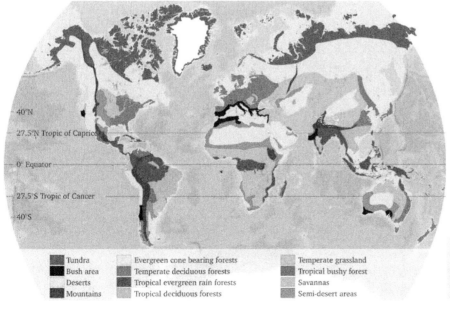

Figure–5.3.: Biomes are the biggest units of ecological systems. Usually biomes are named for their dominant plant species.

BIOMES	CLIMATE	PLANT COVER	ANIMALS	SOIL
Tundra	Polar regions with temperature below 10 C. Here mostly glaciers are present; only three months of the year pass without glaciers.	In the southern regions of the tundra there are small woods, in the northern regions there are grasses.	Fox, reindeer, snowy owls, snow goose, musk ox, and many insect species, but nearly no amphibians or reptiles.	The soil surface in these regions is frozen year-round. This prevents root growth. Plants can grow only where this layer melts.
	This biome has a wet, cool climate. It has higher precipitation than the tundra, with rainfall generally in autumn.	The dominant plant species is pine, spruce, and fir trees.	Deer, bear, falcon, wolf, mouse, bat, squirrel, finches and some insects. Reptiles are scarce, but amphibians are present.	Soil is covered with a thick humus layer. The decomposition of this layer occurs very slowly. The soil is arid and acidic.
	Climate changes from the north to the south. The northern parts are snowy and the soil is frozen. The southern parts are rainy and temperate. The annual rainfall is regular.	At the high altitudes of this biome there are beech, chestnut, linden, and oak trees; near water there are maple, elm, and poplar trees. In the old forests there are shrubs and developed grasses.	Deer, bear, wolf, mountain lion, fox, mouse, wild turkey, woodpecker, and some reptiles, amphibians and insects.	The soil is more productive than taiga. The humus layer on the soil is thin due to the earthworms. The acidity of the soil decreases with depth.
	Cold, wet winters and hot, dry summers. Rainfall is less than in the deciduous forests.	Pastures are dominant in these regions. Secondarily there are flowering and legume plants. There are few trees, mainly near water sources.	Coyote, squirrel, bison, antelope, elk, deer, wolf, puma, grasshoppers, sparrow and insects are present.	Soil is very productive and rich in minerals. Organic substances accumulate at the top of the soil and give it a dark color. This top layer is mostly neutral or basic.
	Heavy rains in winter; summers are hot and arid. (typical Mediterranean climate).	Always covered with shrubs, pine and oak trees are found. The plants here are adapted and resistant to heat and fire.	Deer, mouse, lizards, squirrels and many bird species are present.	The soil layer is thin and unproductive. Fires in the extremely hot summers and falls make the soil dry

 and poor.

BIOMES	CLIMATE	PLANT COVER	ANIMALS	SOIL
Deserts	The temperature is very high during the day and falls suddenly at night. Rainfall is very low, as is moisture.	Spreading bushes are present. Cactus and some other related plants are found as well.	Animals that need little water or store water can live in the desert. Fox, rabbit, antelope, lizards, snakes and some insect species are present.	Since production is very low, organic matter in the soil is low too. The topsoil is alkaline. The soil is rich in minerals except nitrogen.
Tropical Forests	Annual rainfall is high and regular. High temperatures and moisture continue throughout the year.	The flora of the tropical forests is extremely varied. There are trees that grow 25-35 m high. Plants are broad and dark brown colored, thick-leaved and always green.	This biome has a rich fauna as well, including hibernating and migrating animals.	The soil is poor due to the flushing of organic matter and leaching of minerals by rain.
Tropical Shrubs	Hot and arid summers, temperate and wet climate.	Small evergreen trees like oak and maki are present. Since there are frequent fires in this biome only shrub species like manzanita survive.	This biome has a rich fauna as well including hibernating and migrating animals.	The soil is poor due to the flushing of organic matter and leaching of minerals by rain.
Steppe and Savanna	Savannas are rainy from May to October; arid from November to April.	The shrubs are resistant to heat and there are trees shorter than 10 m.	Big cats, mice, antelope, wild cattle are found.	The soil is poor due to the flushing of the minerals.
Semi-desert and Prairie	Prairies are generally seen in the middle of continents with low rainfall and frequent, extreme temperature differences.	Pastures generally have wheat and corn plants. Low rainfall inhibits the growth of tall trees.	Herbivores like bison, deer, and horses.	The prairie soil is deep, productive and rich. In the semidesert, soil is arid and poor.

Figure-5.4.: *Aquatic ecosystem. Since light penetrates to a depth of 100m, this layer is rich in organisms. In the deeper layers live organisms adapted to those conditions.*

Aquatic Ecosystems (Biomes)

In aquatic ecosystems, important environmental factors are salinity, dissolved oxygen, and the availability of light. Aquatic life is ecologically divided into parts.

- Plankton (free-floating)
- Nekton (strong-swimming)
- Benthos (bottom-dwelling)

The microscopic phytoplankton is photosynthetic and is the base of food webs in most aquatic communities. In the sea there is a layering of life zones according to changes in temperature with depth up to 100m below the surface, after which temperature is stable. The zone above this is called the pelagic zone. In every cubic meter of this zone there are millions of microscopic organisms. One of these is phytoplankton, the main source of oxygen. (*Figure-5.4*)

Aquatic biomes are placed in two categories based on salt concentration.

Freshwater ecosystems

Fresh-water ecosystems include rivers and streams (flowing water ecosystems), lakes and ponds (standing water ecosystems), and marshes and swamps (freshwater wetlands). Each type of freshwater ecosystem is distinguished by its own specific environmental conditions and characteristic organisms.

River and Stream Ecosystem: The kinds of organisms found in flowing-water ecosystems vary greatly from one stream to another, depending primarily on the strength of currents. Cold, clean rivers have trout, streams have carp. Because the water in a river flows, it is difficult to classify the fishes. There is great ecological variation between its source (where it begins) and its mouth (where it empties into another body of water). Rivers with cold-water plants are scarce but there are some species of algae and flatworms, frogs and insect larvae.

Lakes and Ponds: Lakes and ponds are standing bodies of water that form in depressions in the earth's crust. They are grouped ecologically into two zones: limnetic (pelagic) and benthic. The **limnetic zone** includes the column of water that fills the depression and covers the **benthic zone**. The organisms found here include **phytoplankton,** blue-green algae, **zooplankton,** fishes, frogs and some insect species. The benthic zone starts at the shoreline and extends to the bottom of the lake. The plants and animals that live in the benthic zone are called **benthos**. Benthos includes water plants, bottom-dwelling organisms like oysters and mussels, worms, and crayfish. The parts of the benthic zones near to the shores **(littoral zone)** have a wide variety of vegetation. In this zone there are plants that rise above the water (reed, cane); plants with leaves that float on the water (lily); and plants that live submerged in the water (elodea).

 Turnover is a mixing of lake waters is called turnover. Fall and spring turnovers cause the mixing of upper and lower layers of water. This circulation, which tends to equalize temperatures throughout the lake, brings oxygen to the oxygen poor depths and minerals to the mineral deficient surface.

 Estuaries form where rivers and streams empty into oceans, mixing freshwater with saltwater. The water in estuaries varies considerably in terms of salinity, temperature and nutrient load. Many species are adapted to estuarine conditions. Tides especially increase the oxygen and nutrients, and organic substances increase the biological diversity. Estuaries are fish nurseries. Many species reproduce there.

There are lots of species of animals as well. In addition, in the sediment on the lake floor there are many protozoa, rotifer and nematode species. Many species of fish live and reproduce here. (*Figure-5.5*)

Lake pollution disrupts the balance of nature. Especially in recent years detergent remaining in wastewater has polluted lakes and harmed the ecological balance. The detergent enriches the water with food substances like nitrogen and phosphorus. This process is called eutrophication. Aquatic plants multiply rapidly and form a large amount of biomass.

The oxygen in the water becomes insufficient to decompose the dead matter. This decreases the water quality. Organisms can't meet their oxygen needs and begin to die, and the lake become useless. This is a kind of water pollution. Another factor that damages the ecological balance of lakes is acid rain. This increases the acidity of the lakes in these regions.

Saltwater (Marine) Ecosystem

Based on ecological features, marine waters are divided into two main zones: **benthic** (ocean floor) and pelagic (ocean water). The benthic zone extends from the shoreline through the ocean floor; **pelagic** zone contains the water column above the benthic zone. Organisms that live in the pelagic zone don't have any interaction with the ocean floor.

Free-living organisms like phytoplankton, zooplankton, cartilaginous and bony fishes, some reptiles, mammals (seals and whales), squid and octopus, shrimp and crab species are some organisms that form the marine ecosystem. (*Figure-5.6*).

There are some organisms that live in the deep ocean where there is no light. These organisms are adapted to such conditions with unique body shapes and feeding styles.

Marine ecosystems are rich in biological diversity. Types of organisms and population size depend on the amount of light, water temperature, pressure, and salinity, currents and tides, as well as the concentration of dissolved minerals and gases and the amount of food in the water.

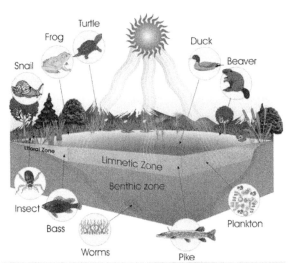

Figure–5.5.: *Layering in lakes and ecologic groupings of organisms living in the lakes. Lakes have benthic and limnetic (pelagic) zones according to the ecologic features.*

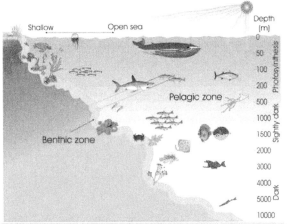

Figure–5.6.: *Marine ecosystem. Seas are ecologically divided into two zones: benthic and pelagic. The benthic zone is sea floor. The pelagic zone is the water column covering the benthic zone. The benthic zone is also layered. Organisms live at various depths within these layers. Organisms live most densely in the top 200m where photosynthesis is most rapid.*

CHOOSE THE CORRECT ALTERNATIVE

1. Which of these is a starting point for primary succession?
   A. on new volcanic island
   B. abandoned farmland
   C. an abandoned city
   D. a fired forest
   E. all of the above

2. All the organisms in a particular area make up a(n) ___ _.
   A. niche   B. food chain   C. population
   D. community   E. ecosystem

3. In an ecosystem, the roles of phytoplankton are _____.
   A. decomposers
   B. producers
   C. primary consumers
   D. secondary consumers
   E. tertiary consumers

4. In the ecosystem, which of the following is an abiotic factor?
   A. Lichen   B. Fungus   C. Algae
   D. Water   E. Blue-green algae

5. Fungi that feed on the remains of plants and animals is acting as a ___ _.
   A. producer
   B. primary consumer
   C. secondary consumer
   D. tertiary consumer
   E. decomposers

6. Which of the following is the largest of Earth's biomes?
   A. Desert   B. Marine   C. Grassland
   D. Freshwater   E. Rain forest

7. Which of the following describes the region where fresh water and salt water mix?
   A. Photic zone   B. Benthic zone
   C. Intertidal zone   D. Aphotic zone
   E. Estuary

8. Which of these biomes is characterized by little rainfall and hot temperature?
   A. Tundra
   B. Tropical rain forest
   C. Taiga
   D. Temperate grassland
   E. Desert

9. Which one of the following biomes is dominated by evergreen trees?
   A. Tropical rain forest
   B. Desert
   C. Deciduous forest
   D. Tundra
   E. Aphotic zone

10. Which of the following is the primary ecological factor determining the distribution of deserts?
    A. Latitude
    B. Windiness
    C. Elevation
    D. Moisture
    E. Temperature

BIOGEOCHEMICAL CYCLES

Chapter 6

BIOGEOCHEMICAL CYCLES

Biogeochemical cycles are the cycling of matter from the environment to living things and back to the environment. Biogeochemical cycles, are also called nutrient cycles that involve both biotic and abiotic components of the ecosystem.

The earth is essentially a closed system (a system from which matter can not escape). The materials are used by organisms can not be lost and it can change its location so materials are re-used and are often re-cycled in the ecosystem.

Four biogeochemical cycles are important for living things.

- Water cycle
- Carbon cycle
- Oxygen cycle
- Nitrogen cycle
- Phosphorus cycle

Carbon, Nitrogen and Water have gaseous forms and they involve atmosphere so cycle over large distances. Phosphorus is an element that is completely non-gaseous form and as a result Phosphorus cycle does not involve the atmosphere, just a local cycling.

 Nutrient circuits involve both biotic and abiotic components of ecosystems and are called biogeochemical cycles.

Water cycle

The water or hydrologic cycle, which continually renews the supply of water that is so essential to life. Water cycle involves an exchange of water between the land, the atmosphere, and living things.

It is assumed that there is nearly 1.4 billion km^3 of water in the world. Though distributed throughout the natural world, most of this water (97%) is in the oceans. Of the Earth's total precipitation (rainfall), 465,000 km3 falls in the sea and 100,000 km^3 falls on land.

There is a strong relationship between the location, duration, and amount of precipitation, and living things. Organisms cannot always use the available water directly, as many factors limit this use. For example, the salinity of seawater and the frozen state of polar water restrict their use by terrestrial organisms. Consequently, living organisms use only 2.6% of the total water mass. At present, rapid population increase and high technology increase the need for water.

The water cycle operates on two physical principles, namely evaporation and condensation. Water absorbs energy and evaporates, and stays in the atmosphere as vapor. As the water vapor rises it collides with cold air currents. The cooled vapor drops back to the earth as rain and snow. Some water falls into the sea, and the cycle begins again.

Underground and aboveground water collects in lakes and seas. From there, as the water warms, it evaporates and enters the air as vapor, and then precipitates again.

> The cycling of water occurs regularly under the influence of sun energy and gravity. The water cycle is a continuous process by which water moves from the earth's surface (lithosphere and hydrosphere) to the atmosphere and back. It is also called the hydrologic cycle. Atmospheric movements and marine currents are important components of the water cycle. The processes of evaporation, condensation and precipitation make up the water cycle.

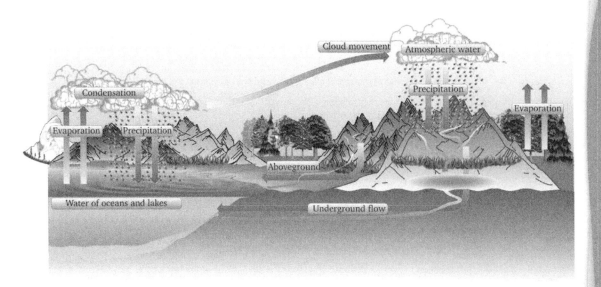

> Carbon dioxide concentrations in aquatic environments are quite different from those on land. Carbon dioxide easily dissolves in water and forms carbonic acid (H_2CO_3), which ionizes to H^+ and HCO_3^-. These ions determine the pH of water.

Carbon cycle

The main source of carbon for organisms is CO_2. Carbon dioxide is found in the lithosphere, hydrosphere, atmosphere and biosphere. Carbon is in the atmosphere as CO_2, in the hydrosphere as bicarbonate ion (HCO_3), in the lithosphere as coal, petroleum, limestone and natural gas, and in the biosphere as the basic raw material of organic substances.

The product of organismal respiration and other sources like forest fires, CO_2 is used in photosynthesis. In respiration, the reverse of this process, organic molecules and O_2 are produced. In other words water and CO_2 are produced from the burning of organic molecules with O_2. Therefore the carbon and oxygen cycles are closely related in nature. The amount of CO_2 in the atmosphere varies from day to night and with the seasons. At night, when photosynthesis stops and all organisms are respiring, the CO_2 level in the atmosphere rises. Likewise, in the seasons when photosynthesis is fast, the CO_2 level in the atmosphere falls. Much research has demonstrated that, because atmospheric CO_2 reduces the reflection of sunlight entering the atmosphere, an increase of CO_2 in the atmosphere results in climatic change, the greenhouse effect.

Saprophytic bacteria and fungi also play a role in returning carbon to the atmosphere. These organisms are essential in the decomposition of dead organisms into inorganic substances. Despite everything, decomposition does not occur completely. Carbon in plant and animal structures is locked into underground reserves through carbonization and petroleum formation. When these formations are extracted and burned as gasoline, natural gas, and coal, CO_2 is released into the environment and used again in photosynthesis.

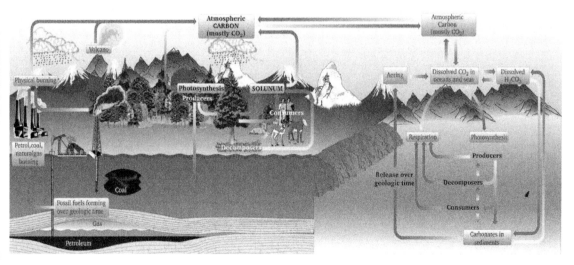

Nitrogen cycle

The nitrogen molecule (N_2), like carbon and oxygen, is an important molecule for organisms. Nitrogen is also a component of molecules like amino acids, nucleic acids, hormones and vitamins. The major sources of nitrogen are the atmosphere and living organisms. The most abundant gas in the atmosphere (78%) is N_2. This atmospheric nitrogen can be used directly by some microorganisms. Plants can use nitrogen in the form of nitrate (NO_3) and ammonium (NH_4) salts. Animals obtain nitrogen from the proteins of the organisms they eat. The cycle of nitrogen between organisms and the atmosphere is a very long and complex process. Actually there are 5-major steps in nitrogen cycle.

Nitrogen fixation: Nitrogen fixing bacteria including cyanobacteria converts atmospheric nitrogen gas (N_2) into ammonia (NH_3) and ammonium (NH_4^+)

Nitrification: Ammonia is converted into nitrate (NO_3^-) by bacteria in the soil known as nitrifying bacteria. Nitrate is the main form of nitrogen absorbed by plants.

Assimilation: Plants use nitrate when they produce protein, nucleic acid and other nitrogen containing compounds, then animals eat plants and nitrogen can pass to animals.

Ammonification: When plants and animals die, the nitrogen compounds in their body are broken down by ammonifying bacteria. And one of the products of this process is ammonia (NH_4^+).

Denitrification: Nitrogen is returned to the atmosphere by denitrifying bacteria, which converts nitrate (NO_3^-) to nitrogen gas (N_2).

> Oxygen is essential for the survival of living things. Oxygen is necessary for respiration and the oxidation of organic substances, and is used in the burning (oxidation) of coal, wood and gas. The atmosphere is 21% oxygen, and 5 % is dissolved in the hydrosphere. The oxygen in nature is produced as a result of photosynthesis. Oxygen also makes up the ozone layer, ozone (O_3) being released as a result of the photolysis of water.

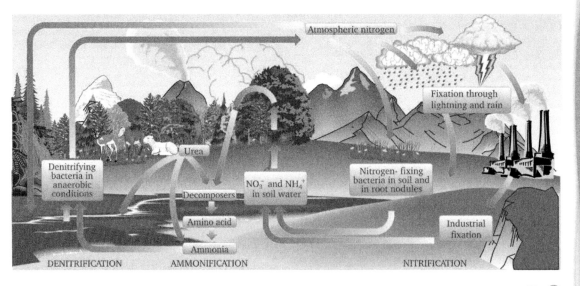

> An ecosystem consists of all the organisms living in a community as well as all the abiotic factors with which they interact. The dynamics of an ecosystem involve two processes: energy flow and chemical cycling. Ecosystem ecologists view ecosystems as energy machines and matter processors. We can follow the transformation of energy by grouping the species in a community into trophic levels of feeding relationships.

Phosphorus Cycle

Phosphorus is another element that is very important for life. Phosphorus is required for the synthesis of nucleic acids, phospholipids and ATP molecules. Moreover it is a component in the structure of the cell membrane, skeleton and skin. The phosphorus cycle is quite different from the nitrogen cycle in that phosphorus does not exist in a gaseous state and therefore does not enter the atmosphere. Phosphorus cycles from land to ocean sediments and back to the land. As water runs over rocks containing phosphorus, it gradually erodes the surface and carriers off inorganic phosphate (PO_4^{-3}) molecules.

The erosion of phosphorus from rocks releases phosphate into the soil where it is absorbed by plant roots. Once inside the plant cells it is converted to organic phosphates. Animals obtain most of their required phosphorus from the food they eat and the water they drink. The remains of dead plants and animals are decomposed to inorganic substances that can be reused by plants. Phosphorus is significant in the efficiency of aquatic and terrestrial habitats. Consequently it is a factor that determines the efficiency of ecosystems.

Certain observations made in oceans show that there is a relationship between fish size, plankton and phosphorus concentration in the water. Phosphate is also mined for agricultural use as phosphate fertilizers. This affects the cycle rate because it speeds up the flow of phosphate from land to sea. Phosphate fertilizers don't remain long in the soil and are carried from the land by streams and rivers to the sea. Erosion caused by human activities, household wastes, and phosphate-containing detergents all increase the flow of phosphates to the seas.

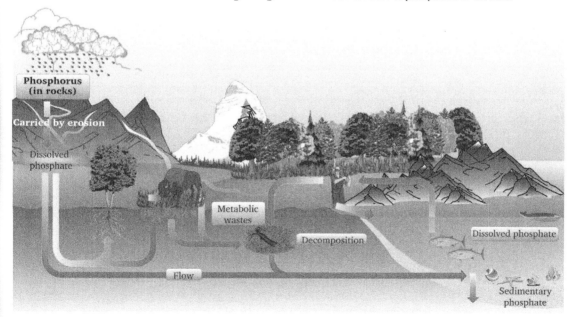

CHOOSE THE CORRECT ALTERNATIVE

1. Which of the following has not evaporation part in its cycle?

   A. Water
   B. Carbon
   C. Nitrogen
   D. Phosphorus
   E. None of the above

2. Which of the following element is fixed from the atmosphere by bacteria to form nitrate?

   A. Phosphorus
   B. Water
   C. Nitrogen
   D. Carbon
   E. Sulfur

3. An ecosystem is unlikely to be limited by the supply of ____ because it is not obtained from the air.

   A. nitrogen
   B. water
   C. carbon
   D. phosphorus
   E. oxygen

4. Nitrifying bacteria convert _____ to _____.

   A. nitrates .....nitrogen gas
   B. nitrogen gas ..... water
   C. nitrogen gas ..... nitrates
   D. ammonium.... nitrites
   E. ammonium.... phosphorus

5. Which of the following process fixes nitrogen from the atmosphere?

   A. Assimilation
   B. Nitrification
   C. Photosynthesis
   D. Denitrification
   E. Nitrogen fixation

6. Recycling of materials are very important for the existence of ecosystems. Which of the following is the primary limiting resource for algae in freshwater lakes?

   A. Carbon
   B. Water
   C. Phosphorus
   D. Nitrogen
   E. None of the above

7. Which of the following is the direction of the global hydrologic cycle supports a net flow of atmospheric water vapor?

   A. From the lakes to the rivers
   B. From the oceans to land
   C. From land to the oceans
   D. From polar to tropical regions
   E. From tropical to polar regions

8. By which process is carbon dioxide is captured from the atmosphere?

   A. Evaporation
   B. Photosynthesis
   C. Cellular respiration
   D. Fermentation
   E. Nitrification

9. Most plants get most of their nitrogen from...

   A. Nitrates in the soil.
   B. N_2 gas in the air.
   C. Proteins.
   D. Ammonium in the soil.
   E. Rainfall.

10. Which of the following biogeochemical cycle in which bacteria play great role?

    A. Nitrogen cycle
    B. Water cycle
    C. Calcium cycle
    D. Phosphorus cycle
    E. Sulfur cycle

ECOLOGY

HUMAN AND THE BIOSPHERE

Chapter 7

ENVIRONMENTAL PROBLEMS

As mentioned previously, every organism is adapted to its living place (habitat). Humans are spread over large areas. In day-to-day life, humans are always interacting with the other living things in their environment. For the continuity of this relationship the ecological balance of the environment must be preserved, but humans frequently use and damage the environment to grow more food, make more shelter and to advance technology.

The environmental system is in balance which ensure the continuity of its living and non-living components, until the second half of 20th century the situation continued in the balance between input matters and output matters, the example of input and output matters are gases, water, salts, energy and different wastes.

But great increase in population, scientific and technological revolution are some modern features of our living century that cause the increasing of natural and manufactured matters that pollute the environment which caused by human activities, the modern economic development, however, sometimes disrupts nature's delicate balance.

Pollution can be defined as the introduction of unwanted or harmful substances into the environment. Pollution caused by human activity has resulted in the extinction of various species of organisms on earth, like the dodo bird and the dusky seaside sparrow.

> Oxygen is essential for the survival of living things. Oxygen is necessary for respiration and the oxidation of organic substances, and is used in the burning (oxidation) of coal, wood and gas. The atmosphere is 21% oxygen, and 5 % is dissolved in the hydrosphere. The oxygen in nature is produced as a result of photosynthesis. Oxygen also makes up the ozone layer, ozone (O_3) being released as a result of the photolysis of water.

Water pollution

Water is one of the most essential necessities of life. All organisms, including humans, need water to live. The hygiene of drinking water is important for health. Factories constructed near rivers and lakes pollute the water. The ecological balance is disturbed. Some organisms die while others carry toxic chemicals in their bodies. Most of the countries are suffering from the pollution of their seas, lakes, rivers, and the running water, which is suitable for daily use. This problem is referred to many reasons:

- Contamination caused by living compounds that cause disease.

- Organic and inorganic compounds that are discharged by factories and house sewerage cause contamination.

- Heat contamination produced by the nuclear- reactor cooling and discharged the factory hot water into the rivers and lakes.

- Kinetic pollution is produced by the movement of boats and ships or from damps.

All the mentioned above cause diminishing (to eliminate) the oxygen rate in the water that effects the well being of all living things in water and encourage the microorganisms like in terrestrial organisms that take oxygen from surrounding water that is the source of many problems and diseases.

The hygiene of fresh water is important for health. Only half of the world's population has access to clean water. Especially in Third World countries, people have to drink water from the places where sewage is dumped. People drinking water from these sources are vulnerable to contagious diseases like cholera, diarrhea, and typhoid.

Soil pollution

Many chemical compounds pollute soil. These pollutants are transform to the soil by irrigation, rain, and wind. Also pollution may occur as a result of using pesticides or from factories waste (gases, radiant, and chemical wastes plastic, metals, wood, paper, packages). They are dissolved in soil and the plants absorb them and then they enters into their tissues. When the animals are fed with such plants, the pollutants will be moved to animal tissues as well. These can be transferred to people as a result of feeding from such plants and meet and dairy food from such animals.

Pollution of soil with agricultural chemicals

Most agricultural chemicals are water-soluble nitrates and phosphates that are applied to fields, lawns and gardens to stimulate the growth of crops, grass and flowers. The chemicals that are used as insecticides include arsenic, mercury and lead, which are highly toxic. Insecticides, since they remain in soil, enter the food chain and poison humans. DDT, which is not biodegradable, and other similar insecticides, accumulates in the fatty tissues of organisms. DDT causes liver cancer, nerve damage, reproductive malfunctions, and death in birds. The effect of DDT is more significant in organisms higher in the food chain.

Herbicidal chemicals, used widely to kill weeds and clear land, also have side effects. America poured 72 million tons of herbicide onto Vietnam to open paths through the jungle during the war from 1961 to 1971. The herbicides, dispersed from airplanes, contained dioxin, a general name for a family of chlorinated hydrocarbons. In the years following the war, high rates of still-birth and premature birth were observed among the Vietnamese. Since similar effects were seen among the American soldiers, the herbicides were investigated. As a result, it was concluded that dioxin causes genetic changes–mutations. At present the use of chemicals containing dioxin is banned.

> The transmission of a toxic substance from one organism to another in a lake ecosystem. These chemicals (e.g. DDT), tranmitted through the food chain but not used in metabolism, accumulate at the end of the chain. Because these chemicals are not metabolized and removed from the tissues, they accumulate in the body. Consequently, the organisms most harmed are those at the end of the food chain.

Air pollution

The tiny layer surrounding the globe is the basic source of air that all living things need and depend on it to carry out their life process. Air contains different gases that they have stable ratios, such as

Oxygen is %21, nitrogen is %78, carbon dioxide is %0.03

Nobel gases is %1 such as (Argon, Helium...etc)

Vapor water that range between %1 in cold and dry air to %4 during humid seasons in the tropical areas.

Any change in the rate of air contents with foreign particles that are contained in air cause the contamination of air.

The Earth is continuously exposed to sunlight that heats the lower layers of the atmosphere. The temperature of the upper atmosphere is lower than the temperature of the lower atmosphere. Air in the lower atmosphere warms and rises, and is replaced by cold air. Accordingly polluted air rises with air currents and spreads all over the world. In this way air pollution from industrialized countries affects other countries, too. Low quality fossil fuels and exhaust released from vehicles are the main sources of air pollution. Though such pollution is temporary, if it stays longer in the air, it may cause death. (*Table-7.1*)

Pollution	Emission	Source
Carbon Monoxide (CO)	100	Motor vehicles and industrial processes
Carbon Dioxide (CO_2)	33	Burning of fossil fuels in power plants
Hydrocarbons	32	Burning of fuels in vehicles and plants
Dust	28	Burning of solid and fluid fluels in plants
Nitrogen oxides (NO, NO_2)	21	Motor vehicles and burning of fluid fuels

Table-7.1.: *The sources of chemicals that cause air pollution and their emission volumes. When the gases present in the composition of the air increase beyond normal levels, all organisms are threatened.*

Figure-7.1.: One of the factors that causes air pollution is CO gas released as a result of fires. Forest fires are the most common of these. Carbon mononoxide gas is released into the environment in vehicle exhaust.

■ **Carbon Monoxide (CO):** Every year 350 million tons of CO is released into the environment. The major source of atmospheric CO is exhaust gases *(Picture-7.1)*. Carbon monoxide is a toxic gas. It binds to hemoglobin strongly in the lungs and prevents the binding of oxygen. The decrease of oxygen transport to the tissues results in headache, lethargy and giddiness. If the concentration of CO exceeds 1%, it is fatal.

■ **Mercury (Hg):** Mercury vapor is released into the air from the burning of coal and gasoline, mining and the smelting of mineral ores. Increased mercury level in the air causes damage to and malfunctions of kidneys and nerves, and death.

■ **Lead (Pb):** Lead vapor, as in the case of mercury, is released into the air by man's modern industrial activities. The main source of lead in the air is exhaust gas. Lead is added to gasoline to increase engine efficiency.

Lead builds up in plants and causes the pollution of food. The symptoms of lead poisoning are giddiness, extreme fatigue, and depression. Lead poisoning causes damage to liver, kidneys and brain.

■ **Chlorofluorocarbons (CFCs):** Chlorofluorocarbons affect the earth's ozone layer. You can learn more about ozone formation and the effects of CFCs on the ozone layer in the "Read Me" section.

Acid Rain

Normal rainwater has very little acid. Acids in the air react with water vapor and form carbonic acid (H_2CO_3). The pH of normal water is around 5.4. Emissions of sulphur dioxide and oxides of nitrogen from power stations, factories, and motor vehicles cause the formation of sulphuric acid and nitric acids in rain clouds. If rain falls through polluted air it picks up more of these gases and increases its acidity. Acidic clouds may be carried away by air currents. When rain falls from acid clouds, it causes a real environmental catastrophe *(Figure-7.2)*. For this reason every country must be sensitive to this issue and take preventive measures. Acid rain is carried from soil to rivers, streams and lakes. The effect of acid rain is greater on the lakes than the rivers and streams. It increases the acidity of the lake water and the ratio of metal salts. As a result, natural life is threatened.

Figure-7.2.: Sulfuric acid and nitric acid are produced from SO_2 and NO_2 gases that are released into the air and mix with water vapor. When this solution falls as acid rain, it causes damage to all organisms and the environment.

Greenhouse effect

Figure: The glass that covers the greenhouse lets the sunlight enter, which warms the air inside. The warm air stays inside and forms a medium where the plants can live and grow faster.

All life processes, the heating of the world, and food synthesis depend directly or indirectly on the sun. Sunlight heats the earth. Some of the light is reflected back by the planet.

Most of the heat reflected from the earth's surface is captured by water vapor in the air. The light that passes through the water vapor is captured by CO_2. Water vapor and CO_2 in the air retain the heat like the glass covering a greenhouse (*Figure-1.60*).

If there were no water vapor and CO_2 in the atmosphere, most of the heat would be reflected and lost to space, resulting in an air temperature of −40 C.

The greenhouse effect is the rise in temperature that the Earth experiences because certain gases in the atmosphere (water vapor, carbon dioxide, nitrous oxide, and methane, for example) trap energy from the sun.

Because of how they warm our world, these gases are referred to as greenhouse gases.

The Earth has warmed by about 0.75C over the past 100 years. Global warming refers to an average increase in the Earth's temperature, which in turn causes changes in climate. A warmer Earth may lead to changes in rainfall patterns, a rise in sea level, and a wide range of impacts on plants, wildlife, and humans.

When scientists talk about the issue of climate change, their concern is about global warming caused by human activities.

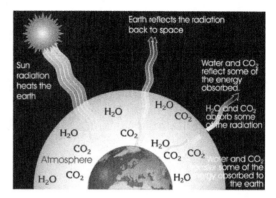

Figure: Some of the light that is reflected by the earth heats water vapor in the atmosphere and some of it is trapped by carbon dioxide. As the density of carbon dioxide in the atmosphere increases, more heat is absorbed, causing a Greenhouse effect on the earth.

The reason for the rise in the CO_2 level in the atmosphere is the burning of fossil fuels and the clearing of the forests. The decrease in the forest cover leads to a decrease in the amount of CO_2 captured by photosynthesis.

Ozone Layer

The word "ozone" comes from the ancient Greek word "ozein", meaning strong-smelling spreading. Ozone is a strong-smelling gas. Ozone has three atoms of oxygen. The ozone layer is located in the stratosphere. Ozone is a blue gas that gives the sky its blue color when the weather is clear. The concentration of ozone there is kept constant by the action of ultraviolet light.

High in the atmosphere, some oxygen (O_2) molecules absorb energy from the sun's ultraviolet (UV) rays and split to form single oxygen atoms. These atoms combine with the remaining oxygen (O_2) to form ozone (O_3) molecules, which very effectively reflect UV rays. The thin layer of ozone that surrounds the earth acts as a shield, protecting the planet from irradiation by UV light. Ozone is present in all the layers of the atmosphere but is concentrated in the stratosphere 12-45 km high. Gases produced on the earth's surface, like CFCs, rise and mix into the stratosphere. Here these gases release Cl atoms.

Chemical reactions begin with the binding of free Cl atoms to the free oxygen atoms. This in turn causes new Cl atoms to be released, which bind to the oxygen atoms. These chemical reactions happen over and over in a cycle. After releasing CFCs into the atmosphere it is impossible to prevent ozone depletion.

The atmosphere protects our world by neutralizing dangerous gases, trapping meteors, and especially by keeping the world warm with a blanket of CO_2.

Formation of the ozone layer and its importance

The major components of this region, by volume, are oxygen (O 21%), nitrogen (N 78%), argon (Ar 0.93%) and carbon dioxide (CO_2 0.033 %). The thickness of the atmosphere is 120 km. The ozone layer covers the outside of the atmosphere. Oxygen molecules are broken down by ultraviolet light and converted into ozone. Oxygen molecules capture UV light of wavelength smaller than 200 nm. UV light of wavelength between 200-300 nm is absorbed by the ozone layer.

Reactions responsible for the formation of ozone:

$$O_2 \xrightarrow{UV} 2O\cdot \quad\quad O\cdot + O_2 \longrightarrow O_3$$

$O + O_2 + M \rightarrow O_3 + M$ (where M indicates other molecules in the medium; conservation of energy and momentum). M removes the excess energy which would otherwise split the O_3 to its constituents.

Reactions responsible for the destruction of ozone:

$$O_3 \xrightarrow{UV} O_2 + O\cdot \quad\quad O\cdot + O_3 \longrightarrow 2O_2$$

Killers of Ozone layer

Chlorine, hydrogen and nitrogen gases are mainly responsible for the destruction of the ozone layer. The most destructive one is Cl. Industry is the origin of 85 % of the ozone-destroying Cl gas. The rest comes from marine reactions.

Three principal gases:

Carbon dioxide (CO_2) (50% of contribution)

Methane (CH_4) (18% of contribution)

Chlorofluorocarbons (CFCs) (14% of contribution)

Manmade sources of carbon dioxide:

1. Fossil-fuel burning
2. Deforestation

Manmade sources of methane:

1. Agro-industrial meat production
2. Intensive agriculture
3. Fermentation and termites

CFCs sources:

1. Aerosol propellants: CFCs such as trichloromethane (CCl_3F) and dichlorodifluoromethane (CCl_2F_2) are normally packed with materials like paint, insecticide or cosmetic preparations in pressurized containers. Upon depressurization by opening the valve, the propellant vaporizes and expels the material inside the can in the form of aerosol spray. Their use is banned and has diminished.

2. Refrigerants: Freon is a series of CFCs, with dichlorodifluoromethane (CCl_2F_2) the most important. Freons absorb the heat of vaporization in evaporation resulting in the cooling of the surroundings. They are widely used as refrigerants in refrigerators and air conditioning units.

3. Foam plastic blowing agent: In making foam plastic, a volatile CFC, trichlorofluoromethane (CCl_3F) is incorporated into the plastic. The heat evolved during the polymerisation reaction vaporises the CFC, which then forms tiny bubbles in the plastic

4. Cleaning solvents: CFCs like trichlorofluoroethane (CCl_2FCClF_2) can dissolve grease and are widely used as solvents in cleaning electric components and metals.

5. Fire extinguishers: Generally contain Br atoms. They are very hazardous to the ozone layer.

 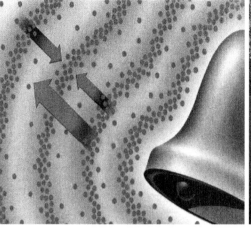 

Noise pollution

Sound is such a common part of everyday life that we often overlook all that it can do. It provides enjoyment, for example, through listening to music or birdsong. It allows spoken communication. It can alert or warn us, say, though a doorbell, or wailing siren. Sound is a part of life. In natural conditions, bird, wind or water sound doesn't disturb us. But the sounds that we call noise disturb humans both physiologically and psychologically. Sound level is measured in decibel (dB). The limits of noise are not certain. But sound between 35-65 dB is psychologically disturbing; between 65-90 dB is peace disturbing; and higher than 90 dB is physiologically harmful noise.

According to its source, noise falls under one of three headings: transport (traffic) noise, industrial noise, and social noise. Transport noise comes mainly from trains, planes, cars, buses, trucks, and motorbikes, and each of these produces noise in a variety of different ways. All of these vehicles make noise because of the friction force between their metal parts and with the air. Social noise includes the noises made by people in parks and at sporting events, as well as radio and TV sounds. Intense noise may rupture the eardrum and cause hearing problems. People living in areas with high levels of noise may experience hypertension, a fast breathing rate, and a high pulse. In addition, noise causes stress, discomfort, anger, and behavioral problems. In noisy workplaces efficiency decreases and attention problems increase. We can control noises by:

- Protecting the human ear by ear covers.

- Using sound insulation protection.

- Eliminating the noises by oiling the machines and using the less noisy machines planning the residential areas away from airports, factories, and high ways.

Noise pollution, human-created noise harmful to health or welfare. Transportation vehicles are the worst offenders, with aircraft, railroad stock, trucks, buses, automobiles, and motorcycles all producing excessive noise. Construction equipment, e.g., jackhammers and bulldozers, also produce substantial noise pollution.

Radiation

Radiation is the process in which energy is emitted as particles or waves. The sun radiates energy continuously. Light coming to the earth is in three groups: ultraviolet light (UV), white light and infrared light. Ultraviolet light has a very small wavelength and high energy level. Therefore, it is dangerous to human health.

The ozone layer reflects most of the UV light before it reaches the atmosphere. Only 2% passes through. An increase in this amount causes certain illnesses like skin cancer. Like solar radiation, underground and underwater deposits of radioactive rock are a natural source of radiation. On the earth uranium (U-235 and U-238), thorium (Th-232), potassium (K-40), strontium (Rb-87) and other radioactive substances are found. When these molecules decay radioactively, energy is emitted. These processes are all natural and have been happening for thousands of years.

Radiation pollution exists because humans use radioactive substances. Modern life, though offering many benefits to humanity, brings many problems. The energy produced from dams and thermal plants was insufficient and people started using nuclear power, the fission of radioactive isotopes, to produce energy. After the discovery of nuclear energy, scientists looked for ways to use it. Nuclear tests were conducted in the Nevada deserts of America, the deserts of Kazakhstan and in the Pacific Ocean by France. In these tests, radiation was released into the atmosphere. Especially in areas near these places the air, water and soil is highly polluted. These radioactive substances are carried away to the other places as well. *(Figure-7.3)*.

Also these countries made ships, submarines and aircraft carriers that run on nuclear energy. They are very efficient economically, but in the event of an accident, malfunction, or technical problem they are a potential threat to the environment and humanity. Nuclear energy is also used to make bombs. Radioactive isotopes are used in medicine and biological research. The substances used in these fields must be handled and disposed of carefully.

Effects of Radiation

Radiation affects the environment both physically and biologically. Nuclear trials and explosions spread dust and smoke which block sunlight. Moreover, the air temperature under the dust layer plunges because the dust blocks the sunlight. As a result there will be serious changes in the climate. The biological effect of radiation is the damage to living things (Figure-1.64). The sensitivity of organisms varies from species to species. For example, insects are more resistant to radiation than birds and mammals. Grasses are more resistant than broad- and needle-leaved plants. *(Figure-7.4)*.

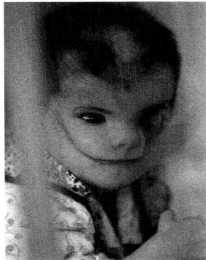

Figure–7.3-4.: *The consequences of the Chernobyll accident have been disastrous. A patient that contracted cancer as a result of the explosion is shown. There are also chronic effects of radiation in addition to its acute effects. As seen in the picture, the genetic structure and development of all organisms living and born in that region are affected. Two significant examples of this are the Hiroshima and Nagasaki cases, and the explosion at Chernobyl.*

Measures against environmental pollution

Some measures that can be taken to reduce the dangers described are given below.

- The waste from factories should be recyclable. In this way we can save the raw materials and also prevent the pollution of the environment

- Fossil fuels used in heating should be high in calories and low in toxic substances so that air pollution is reduced.

- Chimneys of factories and houses, and exhausts of cars should have filtering devices to reduce toxic substances in the air.

- Tree planting should be encouraged so that the gas balance in the atmosphere is maintained and air pollution is reduced.

- Recyclable materials should be collected and used again

- Recyclable materials should be used as much as possible.

- And, most importantly of all, everyone should be trained to be aware of environmental problems.

 Carbon dioxide concentrations in aquatic environments are quite different from those on land. Carbon dioxide easily dissolves in water and forms carbonic acid (H_2CO_3), which ionizes to H^+ and HCO_3^-. These ions determine the pH of water.

> Improvements in the areas of genetics, energy, and chemical synthesis will provide the needed steps for advancements in agriculture and food. Genetically engineered food (which is already on the market) will only increase as scientists and engineers come to a better understanding of genetics. More efficient energy sources may allow for easier food production locally as well as cheaper methods of harvesting and allow for cheaper irrigation and water pumping. Chemical synthesis may allow for better pesticides, fertilizers, and even soils for growing crops.

Humans and environment

Humans, like other organisms, are participants in the food chain and part of the balance of nature. Like other heterotrophs they eat, excrete and need heat and shelter. All of these necessities are met by nature. However, the technology developed by humans for a better and more comfortable life can adversely affect the environment and may damage the ecological balance on earth. The advantages and disadvantages of agriculture will be discussed in the following articles. Humans also cut and process trees to make houses, furniture, paper and decorative products. As a result of the timber industry and fires, forests are destroyed, causing air pollution and erosion. The activities of humans sometimes damage the ecological balance of nature. Industrialization produces benefits for people, and problems for the ecological balance.

Agriculture

The rapidly growing world population has brought food problems to the agenda. New methods in agriculture and stock-breeding are developed and used to produce food more abundantly. In cereal-grain agriculture, artificial fertilizers are used to produce a larger harvest in a shorter time. Insecticides are used against insects and herbicides against weeds to kill them and protect the crop. The use of technology to increase productivity is called intensive agriculture. Scientists and engineers have developed new systems to reduce the costs of intensive agriculture. The side effects of these methods are not understood until later. In stock-breeding, animals are fed special foods in a closed place. Since the animals move very little and gain weight, they are butchered sooner. Fattening livestock for market increases the cost of their meat.

Irrigation

One of the most important requirements of agriculture is water. Especially in arid places, it rains at some times of the year and it is dry at others. Dams are built to save water in arid places. Water collects behind the dams and is used for irrigation. With the proper use of irrigation, a larger harvest is possible. Despite the advantages of irrigation, excess irrigation brings some dangers to the environment.

- Irrigation drains minerals away from the soil through leaching.

- Salt present deep in the soil dissolves in the water and rises up, causing the soil to become salty and dry.

- Microorganisms carrying contagious diseases are spread through irrigation canals.

- Freshwater fish and, in turn, the food chain are affected adversely.

Figure–7.5.: *Regular irrigation increases the quality and efficiency of agriculture, but irregular irrigation causes more problems than it solves, even causing aridity of the soil and reducing efficiency.*

The Effects of Mechanized Agriculture on Natural Life

The transition from simple agricultural tools to modern agricultural methods started with the Industrial Revolution. The use of modern agricultural machines has decreased the human effort required, while the quality and performance of the product has increased. Though there are a lot of advantages to using agricultural machines, there are some disadvantages as well. Some of these are stated below (*Figure-7.6*).

- Since modern agricultural devices cannot be used in narrow fields, these fields were widened by cutting trees. Some species were destroyed or became extinct as a result.

- The weight of the machines compresses the soil and decreases its permeability for water, causing the accumulation of water above the soil.

- The use of modern agricultural machines increased the costs and the sale prices.

- With the removal of ecological boundaries, natural events like wind and flood cause erosion and evaporation of excess water..

Figure–7.6.: *Certainly agricultural tools have reduced the work of people and increased their efficiency. When these tools are used according to regulations, soil erosion and excess evaporation can be prevented. There are, however, disadvantages to using these tools.*

Monocultivation (Growth of the same product every year)

Since every agricultural product has a different price in the market, farmers tend to grow the crops that bring more money. Accordingly, farmers may grow the same crop every year. The advantage of growing the same crop every year is the ability to use the same machines. In this way the cost of mechanization and labor falls.

The disadvantages of monocultivation are: Since the same species is grown every year the same minerals are removed excessively, which causes the soil to become arid. Though this condition is compensated for by the application of artificial fertilizers to the soil, the cost increases. The repeated growth of the same species boosts the insect population.

Artificial fertilizers

In this century, the use of artificial fertilizers started the green revolution. In most parts of the world arid areas are converted to agriculture fields. At the same time the productivity of the existing fields is increased. The elements C, N, O, P, H and K are very essential in the structure of organic molecules. Plants obtain C, H and O from air and N, P and K from the soil. The abundance of these minerals in the soil increases the quality of the products. Most of the fertilizers used in the world are of the N–P–K variety. In 1990, 20 million tons of NPK fertilizer were used around the world. The use of fertilizers increases not only the quality but also the costs, so it must be used efficiently. For timing and dosage of fertilizer application, farmers should consult experts.

Humans, like other organisms, are participants in the food chain and part of the balance of nature. Like other heterotrophs they eat, excrete and need heat and shelter. All of these necessities are met by nature. However, the technology developed by humans for a better and more comfortable life can adversely affect the environment and may damage the ecological balance on earth.

Though it is advantageous to use fertilizers there are some drawbacks as well.

The benefits of artificial fertilizers:

- The use and storage of artificial fertilizers is easier.
- The type and dose of the minerals can be adjusted according to the needs of the soil.
- The soil can be planted and even provide more than one crop per year because it does not need to be left fallow.

The drawbacks of artificial fertilizers:

- Since artificial fertilizers do not contain humus, the quality of the soil decreases gradually, becoming more susceptible to erosion.
- Excess use of fertilizers is a waste of money and energy.
- Rain leaches soluble substances from the soil. Excess fertilizer is carried away to rivers and lakes. The transport of nitrate and phosphate fertilizers to lakes and rivers causes the destruction of natural life. If the nitrate solution seeps into drinking water, it may threaten human life. The cultivation of crops like wheat and barley enables the efficient use of fertilizers without waste.
- Artificial fertilizers disrupt the structure and texture of the soil, gradually making it more difficult to cultivate.
- Since the mineral concentration increases in the soil, plants can't absorb water through their roots. This condition may cause the death of the plant.

Fertilizers typically provide, in varying proportions, the three major plant nutrients (nitrogen, phosphorus, and potassium), the secondary plant nutrients (calcium, sulfur, magnesium), and sometimes trace elements (or micronutrients) with a role in plant nutrition: boron, manganese, iron, zinc, copper and molybdenum.

READ ME!
An ecological disaster

Today everybody is aware that nature is gradually losing its balance due to exploitation. This is mostly as a result of human selfishness. Especially today people unknowingly sacrifice nature for the sake of technological advances. Pollution of the environment is increasing rapidly. Though the pollution of soil, water and air continues to increase, still no precautions are being taken to stop it. No steps were taken in advance to save the ozone layer and, in the same way, no effort is being made to stop the Greenhouse effect.

On the contrary, necessary measures should be taken at once, because if measures are not taken to prevent pollution, humanity and other organisms are going to suffer.

Substances like CO_2, heat energy and wastes are released into the environment and disrupt the balance of nature.

According to research, every year as a result of the respiration of humans and animals 92 billion tons of CO_2, and from the respiration of plants, 37 billion tons of CO_2 is released into the atmosphere. Moreover, the burning of fossil fuels in factories and household heating systems releases 18 billion tons of CO_2. In total 147 billion tons of CO_2 is sent into the environment. As can be seen the amount of CO_2 in the atmosphere increases continuously.

The only organisms that assimilate CO_2 from the atmosphere and synthesize organic molecules, and in turn release O_2 to the atmosphere, are the photosynthetic organisms. According to today's data, green plants assimilate 129 billion tons of CO_2 from the atmosphere. The 18 billion ton disparity is decreased with the CO_2 - O_2 cycle. But the balance is still not maintained. Especially with the increase of air, land and water pollution, the maintenance of the balance becomes impossible.

Transport vehicles, factories, and heating systems release trillions of kilocalories of heat energy. Green plants that keep the balance of CO_2 and O_2 in the air also maintain the heat balance in nature. Among living things only green plants are endothermic and absorb heat energy from the outside, producing organic compounds not only for themselves but also for other organisms. These organisms fix 675 kilocalories of energy into every glucose molecule. Generally the annual plant production on earth is 130 billion kilocalories on land and 30 billion kilocalories in oceans. To produce that much energy, plants take in trillions of kilocalories of energy from the environment.

To compensate for the effects of technological advances, the green cover of the earth should be increased. Consequently, to make the world livable again, we should arrange campaigns for planting trees and conserving nature.

Adapted from Bilim ve Teknik Dergisi.

Use of Herbicides in Biological Struggle

Mineral-rich soil is a suitable place for weeds as well as crops. Weeds compete with crop plants for light, water and minerals. The simplest method of fighting them is to dig them up. Since this takes too much time and effort, chemicals are used to kill them. The chemicals, called herbicides, kill the weeds and save the crops. The herbicides target the broad-leaved plants to kill. Small-leaved crops are not killed by the herbicides. Not all herbicides are selective–some of them kill other organisms as well. The use of selective herbicides is one of the inevitable methods of modern agriculture. Otherwise the amount and quality of the harvest will decrease. Insects, weeds and disease destroy one-third of the world's total agricultural production. Insects rank first among these. For example, weevils are a real nuisance in wheat fields. Potato beetles attack tuberous plants like potatoes. Insects are killed with certain chemicals called insecticides. There are also beneficial insects that provide pollination. They must be saved while killing the others.

Use of DDT in Agriculture

The most commonly used insecticide is DDT. DDT was discovered by Paul Mueller in 1939 who observed that it kills many insects including flies. The World Health Organization (WHO) stated that DDT killed many insects and prevented diseases and the destruction of crops, saving the lives of five million people. These are the benefits of DDT. Its drawbacks were explained in the soil pollution section.

READ ME!
An ecological disaster

The use of chemicals against pests and weeds increased in the Canate prairie in Peru after 1949. Cotton production increased dramatically in the following years. Therefore, farmers used more chemicals to get a bigger harvest, and they sprayed the chemicals more frequently. To spray easily from planes they chopped all the trees and shrubs. After 1965 the worms that damage cotton increased and started to reproduce massively and that year was a catastrophe for cotton production.

We can explain the situation ecologically.

- The chemicals that destroy cotton worms also killed the insect predators of these worms.
- With the destruction of trees and shrubs the habitat of birds and insects was destroyed as well.
- The cotton worms became resistant to the chemicals. Since their predators were destroyed, their population naturally increased quickly.
- As a conclusion, the removal from the food chain of the insects that feed on these worms played an important role in this disaster.

Erosion and forests

The richness of the composition of the soil is very important for the growth of plants. Soil, a thin layer on the uppermost part of the earth's surface, is enriched by the activities of decomposers that break down the remains of dead plants and animals. But rivers and floods wash away this productive layer. Trees that block and lower the speed of the water, and hold the soil with roots protect the soil layer.

A feature of forests more important than soil conservation is the maintenance of the oxygen balance of the earth. The thoughtless and negligent cutting of forests, the oxygen tank of organisms, threatens the future of all living things.

Energy

Energy is an inevitable part of our lives. Transportation, heating, illumination and other activites like the use of houshold electrical devices demand electricity. Excess dependence on electricity has forced scientists to look for alternative energy sources. Countries prepare and apply their energy policies according to the needs of their economies and their reserves of natural resources. Countries with many rivers produce hydro-electricity. Even though dams cause the least damage to the environment, they alter the natural water cycle and floodplain which consequently results in climate change. As a result of climate change, some species decrease in number or disappear, while other species reproduce more and increase in number. The reclamation of the area around the dam takes a long time.

Some countries use coal, natural gas, oil and other fossil fuels to produce electricity. The smoke and gases that are released as by-products of the burning of fossil fuels pollute the air. Consequently natural life is affected and ecosystems change.

> A nuclear power plant (NPP) is a thermal power station in which the heat source is one or more nuclear reactors generating nuclear power.
>
> Nuclear power plants are base load stations, which work best when the power output is constant (although boiling water reactors can come down to half power at night). Their units range in power from about 40 MWe to over 1000 MWe. New units under construction in 2005 are typically in the range 600-1200 MWe.

> Over-application of chemical fertilizers, or application of chemical fertilizers at a time when the ground is waterlogged or the crop is not able to use the chemicals, can lead to surface runoff (particularly phosphorus) or leaching into groundwater (particularly nitrates). One of the adverse effects of excess fertilizer in lacustrine systems are algal blooms, which can lead to excessive mortality rates for fish and other aquatic organisms.

The most dangerous but the cheapest and most effective energy source is nuclear energy. Many countries use energy produced in nuclear power plants, especially countries without reserves of fossil fuels and water. When they work without any problems, nuclear power plants are very economical, but they are a big potential threat to the environmen.

For example, the Chernobyl nuclear plant was providing electricity to a certain region of the Ukraine. An exercise conducted carelessly caused the reactor to explode. This disaster adversely affected the Ukraine and the effects are still visible. The explosion and dispersion of radiation affected other countries as well.

Industrialization

Natural life was preserved and the products and activities of man did not harm the environment until the end of the 1800s. The rapid development of technology as a result of the industrial revolution, which made life easier for humans in many respects, brought with it its own problems. The release of CO_2 from the chimneys of factories caused global warming. The nitrogen and sulphur exhaust from some factories caused air pollution.

Moreover, man-made substances remain in the environment for a long time and don't recycle for many years, if ever. The release of these products into the environment brings water and soil pollution.

The pollution that comes with industrialization doesn't affect only the industrialized countries but also affects other, under-developed countries as well.

> Nuclear power plants are classified according to the type of reactor used. However some installations have several independent units, and these may use different classes of reactor. In addition, some of the plant-types below in the future may have passively safe features.
>
> **Fission reactors:** Fission power reactors generate heat by nuclear fission of fissile isotopes of uranium and plutonium.
>
> **Fusion reactors:** Nuclear fusion offers the possibility of the release of very large amounts of energy with a minimal production of radioactive waste and improved safety. However, there remain considerable scientific, technical, and economic obstacles to the generation of commercial electric power using nuclear fusion.

Setting up and using a Terrarium

To examine a natural population in an ecosystem, you may set up a model population in the laboratory. In order to make such a model work, you should be informed about the organisms and what kind of environment they live in and which organisms they interact with. A terrarium is like an aquarium prepared for terrestrial organisms.

You can get a glass container by buying one from a supplier, or you can make one yourself. The top opening of the terrarium must be closed with a *flywhisk*. You can put different populations into the terrarium. We will mention some of these here.

Preparing a terrarium containing a decaying log

For this you need to find a decaying log. You may cut the log to a suitable size for your terrarium using an axe. It is not advised to use soil.

There are many organisms living on the decaying log, such as ants, bacteria, some plants, spiders, centipedes and some other insects. Among these organisms there are eggs and larvae as well. You can easily observe the growth and development of these eggs and larvae into adults.

To feed the organisms put some breadcrumbs and a wet sponge in the terrarium. Add some water to keep the terrarium moist. Try to identify the species of organisms and their roles in the population.

Preparing a terrarium containing a desert populationPut a large amount of sand (white sand) and a few pieces of rock into your terrarium. In this condition, animals or plants or both can live together. You should add plants adapted to desert conditions and storing water (e.g. cactus); and of animals you may put insects, spiders, and lizards. To feed the lizard you may add earthworms and insects, and don't forget to add a water bowl.

Discuss how these plants are adapted to these conditions.

Preparing a terrarium containing populations of forest-floor organisms

Put a lot of soil onto the floor of the terrarium. Press the soil in a certain area and add some water. You may cover the surface of the soil with mosses and ferns, and plant tree seedlings. You may even add some mushrooms.

If fed, land or water frogs can live in this condition as well. Think of other organisms that you may add to this system.

Preparing an aquarium containing populations of aquatic organisms

Put some sand and small sea stones in the aquarium. Some ornamental objects, snail shells, and different shaped stones may be used to beautify the aquarium.

You should prepare the necessary conditions for the survival of the organisms that you are going to add. The temperature of the water, cleanliness, oxygen content, and places for shelter and reproduction must all be considered.

Therefore, an automatic heater, aeration tubes and pump, filter, thermometer, and some chemicals may be obtained. After preparing the environment with these things you may add some plants and algae. You may also add some fish species, frogs, aquatic turtles and snakes.

Determine the food chain among the organisms in the aquarium.

CHOOSE THE CORRECT ALTERNATIVE

1. What is Ozone?
   A. A harmful gas in the upper atmosphere.
   B. A beneficial gas in the upper atmosphere.
   C. A beneficial gas in the lower atmosphere.
   D. A gas which is harmful in the upper atmosphere and beneficial in the lower. atmosphere
   E. A gas which is beneficial in both the upper atmosphere and the lower atmosphere.

2. Which of the following is a greenhouse gas that may contribute to global warming?
   A. Methane   B. Carbon dioxide
   C. Phosphorus   D. Oxygen
           E. Nitrogen

3. More fuel-efficient cars produce less CO_2, which might reduce...
   A. Ozone depletion.
   B. Global warming.
   C. Acid precipitation.
   D. Bioremediation.
   E. All of the above

4. Which of the following is the major cause of extinction of species today?
   A. The greenhouse effect
   B. Habitat destruction
   C. Global warming
   D. Acid precipitation.
   E. Ozone depletion.

5. About how much of the Earth's surface is covered by water?
   A. 50%   B. 60%   C. 70%   D. 80%   E. 90%

6. Most of Earth's water is contained in:
   A. rivers
   B. polar ice
   C. atmosphere
   D. oceans
   E. lakes

7. The Earth's total water supply is about:
   A. 510 million cubic miles
   B. 320 million cubic miles
   C. 150 million cubic miles
   D. 80 million cubic miles
   E. 550 million cubic miles

8. About what percentage of Earth's water is salt water?
   A. 33%   B. 52%   C. 83%   D. 97%   E. 99%

9. About how much of all of Earth's water is usable by humans?
   A. 3%   B. 3%   C. 12%   D. 25%   E. 69%

10. Which one of the below does NOT belong to the group of organisms called plankton?
    A. Krill
    B. Jelly fish
    C. Squid
    D. Diatoms
    E. Copepods

APPENDIX

Ecology

SCIENTIFIC MEASUREMENT

The relationship between mass and volume of water

(at 20°C)

1 g = 1 cm³ = 1 mL

Some Common Units of Length

Unit	Abbreviation	Equivalent
meter	m	39 inches
centimeter	cm	0.01 meter
millimeter	mm	0.001 meter
micrometer	mm	10^{-6} (one-millionth) of a meter
nanometer	nm	10^{-9} (one-billionth) of a meter
angstrom	A	10^{-10} (one-trillionth) of a meter

dalton or atomic mass unit (amu) the approximate mass of a proton or neutron

mole the formula weight of a substance expressed in grams

Avogadro's number (N)

6.02×10^{23} the number of particles in one mole of any substance

Some Common Units of Mass

Unit	Abbreviation	Equivalent
kilogram	kg	1000 gram
gram	g	10^{-3} kg
milligram	mg	10^{-3} gram
microgram	mg	10^{-6} gram
nanogram	ng	10^{-9} (one-billionth) of a gram
picogram	pg	10^{-12} (one-trillionth) of a gram

Energy Conversions

calorie (cal) energy required to raise the temperature of 1 g of water (at 16°C) by 1°C

1 calorie = 4.184 joules

1 kilocalorie (kcal) = 1000 cal

Some Common Units of Volume

Unit	Abbreviation	Equivalent
liter	L	1000 milliliter
milliliter	mL	10^{-3} L (1 mL = 1 cm³ = 1 cc)
microliter	mL	10^{-6} Liter

GLOSSARY

A

Abiotic: Non-living; usually applied to the physical and chemical aspects of an organism's environment.

Abundance: The number of organisms in a population, combining 'intensity' (density within inhabited areas) and 'prevalence' (number and size of inhabited areas).

Acid rain: Rain with a very low pH (often below 4.0) resulting from emissions to the atmosphere of oxides of sulfur and nitrogen.

Adaptation: Inherited characteristics that enhance the ability of an organism to survive and reproduce in a particular environment.

Aestivation: A state of dormancy during the summer or dry season.

Age structure: The relative number of individuals of each age in a population.

Aggregated distribution: The distribution of organisms in which individuals are closer together than they would be if they were randomly or evenly distributed.

Amensalism: An interaction in which one organism (or species) adversely affects a second organism (or species), but the second has no effect (good or bad) on the first.

Angiosperms: Flowering plants. Strictly, those seed-bearing plants that develop their seeds from ovules within a closed cavity.

Annual: A species with a life cycle which takes approximately 12 months or rather less to complete, whose life cycle is therefore directly related to the annual cycle of weather, and whose generations are therefore discrete.

Aphotic zone: The part of the ocean beneath the photic zone, where light does not penetrate sufficiently for photosynthesis to occur.

Arthropod: A member of the animal phylum Arthropoda, which includes the insects, crustaceans (e.g. crabs, shrimps, barnacles), spiders, scorpions, mites, millipedes and centipedes.

autotrophs: Organisms that synthesize their own nutrients; include some bacteria that are able to synthesize organic molecules from simpler inorganic compounds.

B

Behavior: What an animal does and how it does it.
Behavioral ecology: A heuristic approach based on the expectation that Darwinian fitness (reproductive success) is improved by optimal behavior.

Benthos-Benthic communities: The plants, microorganisms and animals that inhabit the bed of aquatic environments.

Biodiversity: In its most general sense, biodiversity refers to all aspects of variety in the living world. Specifically, the term may be used to describe the number of species, the amount of genetic variation or the number of community types present in an area.

Biogeochemical cycling: The movement of chemical elements between organisms and non-living compartments of atmosphere, lithosphere and hydrosphere.

Biogeography: The study of the geographical distribution of organisms.

Biological clock: An internal timekeeper that controls an organism's biological rhythms; marks time with or without environmental cues but often requires signals from the environment to remain tuned to an appropriate period.

Biological control: The use of a pest's natural enemies in order to control that pest.

Biological oxygen demand: The rate at which oxygen disappears from a sample of water - a measure of deoxygenating ability commonly used as an index of the quality of sewage effluent.

Biological pesticides: A preparation used to provide immediate control of a pest, and which consists of biological as opposed to chemical material.

Biomagnification: The increasing concentration of a compound in the tissues of organisms as the compound passes along a food chain, resulting from the accumulation of the compound at each trophic level prior to its consumption by organisms at the next trophic level.

Biomass: The weight of living material. Most commonly used as a measure per unit area of land or per volume of water.

Biome: One of the major categories of the world's distinctive plant assemblages, e.g. the tundra biome, the tropical rainforest biome.

Biotic: Living; usually applied to the biological aspects of an organism's environment, i.e. the influences of other organisms.

Bioremediation: The use of living organisms to detoxify and restore polluted and degraded ecosystems.

C

C3 plant: Plant that carries out carbon fixation solely by the Calvin cycle.

C4 plant: Plant that fixes carbon initially by the Hatch-Slack pathway, in which the reaction of CO_2 with phosphoenolpyruvate is catalyzed by PEP carboxylase in leaf mesophyll cells; the products are transferred to the bundle sheath cells, where the Calvin cycle takes place.

Calorie: The amount of heat energy required to raise the temperature of 1 g of water 1°C; equivalent to 4.184 joules.

Canopy: The uppermost layer of vegetation in a terrestrial biome

Carnivore: The consumption by an organism of living animals or parts of living animals.

Carrying capacity: The maximum population size that can be supported indefinitely by a given environment.

Chaparral: A thicket of low evergreen oaks or dense tangled brushwood.

Climate: The prevailing weather conditions at a locality.

Climax: The presumed endpoint of a successional sequence; a community that has reached a steady state.

Clumped distribution: The distribution of organisms in which individuals are closer together than if they were distributed at random or equidistant from each other.

Colonization: The entry and spread of a species (or genes) into an area, habitat or population from which it was absent.

Commensalism: A symbiotic relationship in which the symbiont benefits but the host is neither helped nor harmed.

Community: The species that occur together in space and time. All the organisms that inhabit a particular area; an assemblage of populations of different species living close enough together for potential interaction.

Competition: An interaction between two (or more) organisms (or species), in which, for each, the birth and/or growth rates are depressed and/or the death rate increased by the other organisms (or species).

Competitive exclusion: The elimination from an area or habitat of one species by another through interspecific competition. The concept that when populations of two similar species compete for the same limited resources, one population will use the resources more efficiently and have a reproductive advantage that will eventually lead to the elimination of the other population.

Conservation: The principles and practice of the science of preventing species extinctions.

Continental drift: The separation and movement of land masses in geological time.

D

Decomposers: Any of the saprotrophic fungi and bacteria that absorb nutrients from nonliving organic material such as corpses, fallen plant material, and the wastes of living organisms, and convert them into inorganic forms.

Decomposition: The breakdown of complex, energy-rich organic molecules to simple inorganic constituents.

Density: The number of individuals per unit area or volume.

Density dependence: The tendency for the death rate in a population to increase, or the birth or growth rate to decrease, as the density of the population increases.

Density independence: The tendency for the death, birth or growth rate in a population neither to rise nor fall as the density increases.

Desert: A desolate and barren region, usually deficient in available water, and with scant vegetation.

Detritivory: Consumption of dead organic matter (detritus) usually together with associated microorganisms.

Disease: The disturbed or altered condition of an organism (malfunctioning) caused by the presence of an antagonist (toxin or pathogen) or the absence of some essential (e.g. micronutrient or vitamin).

Dispersal: The spreading of individuals away from each other, e.g. of offspring from their parents and from regions of high density to regions of lower density.

Disturbance: A force that changes a biological community and usually removes organisms from it. Disturbances, such as fire and storms, play pivotal roles in structuring many biological communities.

Dominant species: Species which make up a large proportion of community biomass or numbers.

Dormancy: A condition typified by extremely low metabolic rate and a suspension of growth and development. An extended period of suspended or greatly reduced activity, e.g. aestivation and hibernation.

Dynamic equilibrium: The state of a system when it remains unchanged because two opposing forces are proceeding at the same rate.

E

Ecological niche: A term with alternative definitions, not all of them synonymous. To state two:(i) the 'occupation' or 'profession' of an organism or species; or (ii) The sum total of a species' use of the biotic and abiotic resources of its environment.

Ecology: The study of how organisms interact with their environments.

Ecosystem: All the organisms in a given area as well as the abiotic factors with which they interact; a community and its physical environment.

Ectoparasites: Parasites that feed on the external surface of a host.

Ectotherm: An animal, such as a reptile, fish, or amphibian, that must use environmental energy and behavioral adaptations to regulate its body temperature.

Effective population size: The size of a genetically idealized population with which an actual population can be equated genetically.

Emigration: The movement of individuals out of a population or from one area to another.

Endemic: Having their habitat in a specified district or area, or the presence of a disease at relatively low levels, all the time.

Endotherm: An organism which is able to generate heat within itself to raise its body temperature significantly.

Epidemic: The outbreak of a disease which affects a large number and/or proportion of individuals in a population at the same time.

Epiphyte: A plant that nourishes itself but grows on the surface of another plant for support, usually on the branches or trunks of tropical trees.

Estivation: A physiological state characterized by slow metabolism and inactivity, which permits survival during long periods of elevated temperature and diminished water supplies.

Estuary: The area where a freshwater stream or river merges with the ocean.

Eutrophication: Enrichment of a water body with plant nutrients; usually resulting in a community dominated by phytoplankton.

Evapotranspiration: The water loss to the atmosphere from soil and vegetation. The potential evapotranspiration may be calculated from physical features of the environmental such as incident radiation, wind speed and temperature. The actual evapotranspiration will commonly fall below the potential depending on the availability of water from precipitation and soil storage.

Exponential growth: Growth in the size of a population (or other entity) in which the rate of growth increases as the size of the population increases.

Extinction: The condition that arises from the death of the last surviving individual of a species, group, or gene, globally or locally.

F

Fitness: The contribution made to a population of descendants by an individual relative to the contribution made by others in its present population. The relative contribution that an individual makes to the gene pool of the next generation.

Food chain: An abstract representation of the links between consumers and consumed populations, e.g. plant - herbivore - carnivore. The pathway along which food is transferred from trophic level to trophic level, beginning with producers.

Food web: Representation of feeding relationships in a community that includes all the links revealed by dietary analysis.

fossil fuels: Energy deposits formed from the remains of extinct organisms; fossil fuels include coal, oil, and natural gas.

G

Gamete: Haploid reproductive cells (ovum and sperm).

Gene: A unit of inherited material - a hereditary factor.

Global warming: The predicted warming of the planet resulting from increasing atmospheric concentrations of radiative gases such as carbon dioxide, methane, nitrous oxide and chlorofluorocarbons.

Greenhouse effect: The heating that occurs when gases such as carbon dioxide trap heat escaping from the Earth and radiate it back to the surface; so-called because the gases are transparent to sunlight but not to heat and thus act like the glass in a greenhouse.

H

Habitat: Place where an oorganism lives.

Habituation: A very simple type of learning that involves a loss of responsiveness to stimuli that convey little or no information.

Halophyte: A plant that tolerates very salty soil.

Haustoria: Branches of parasitic plants or fungi which enter the tissues or cells of the host.

Hemiparasite: Plants which are photosynthetic but form connections with the roots or stems of other plant species, drawing most or all of their water and mineral nutrient resources from their host.

Herbicide: A chemical or biological preparation which kills plants.

Herbivore: A heterotrophic animal that eats plants.

Heterotroph: An organism that obtains organic food molecules by eating other organisms or their by-products.

Hibernate: To remain dormant during the winter period. A physiological state that allows survival during long periods of cold temperatures and reduced food supplies, in which metabolism decreases, the heart and respiratory system slow down, and body temperature is maintained at a lower level than normal.

Holoparasite: Parasitic plants which lack chlorophyll and are therefore wholly dependent on their host plant for the supply of water, nutrients and fixed carbon.

Homeotherm: An organism which maintains an approximately constant body temperature, usually above that of the surrounding medium.

Host: An organism which is parasitized by a parasite.

Hydrological cycle: The movement of water from ocean, by evaporation, to atmosphere, to land and back, via river flow, to ocean.

Hydrophilia: An overwhelming desire for water.

I - J

Immigration: Entry of organisms to a population from elsewhere.

Interspecific competition: Competition between individuals of different species.

Intraspecific competition: Competition between individuals of the same species.

Isotherm: A line on a map that joins places having the same mean temperature.

joule (J): A unit of energy: 1 J 50.239 cal; 1 cal 54.184 J.

K

K selection: The concept that in certain (K-selected) populations, life history is centered around producing relatively few offspring that have a good chance of survival. A small reproductive allocation, much parental care, and the production of few but large offspring.

Kilocalorie: The energy needed to heat 1000 grams of water from 14.5 to 15.5 degrees C.

L

Landscape: Several different primarily terrestrial ecosystems linked by exchanges of energy, materials, and organisms.

Learning: A behavioral change resulting from experience.

Life cycle: The sequence of stages through which an organism passes in development from a zygote to the production of progeny zygotes.

Life form: Characteristic structure of a plant or animal.

Life history: An organism's lifetime pattern of growth, differentiation, storage and reproduction.

Littoral zone: The zone at the edge of a lake or ocean which is periodically exposed to the air and periodically immersed.

M

Macronutrients: 1. Elements needed by plants in relatively large (primary) or smaller (secondary) quantities. 2. Foods needed by animals daily or on a fairly regular basis.

Maturation: The process of becoming fully differentiated, fully functional and hence fully reproductive.

metabolism: The sum of all the chemical processes that occur within a cell or organism: the transformations by which energy and matter are made available for use by the organism.

Microbes: Microorganisms: any microscopic organism, including bacteria, viruses, unicellular algae and protozoans, and microscopic fungi such as yeasts.

Microclimate: The climate within a very small area or in a particular, often tightly defined, habitat.

Micronutrient: An element that an organism needs in very small amounts and that functions as a component or cofactor of enzymes.

Migration: The movement of individuals, and commonly whole populations from one region to another.

Monoculture: Cultivation of large land areas with a single plant variety. A large area covered by a single species (or, for crops, a single variety) of plant; or, in experiments, plants of the same species grown alone without any other species.

Mutualism: A symbiotic relationship in which both participants benefit. An interaction between the individuals of two (or more) species in which the growth, growth rate and/or population size of both are increased in a reciprocal association.

Mycorrhiza: A mutualistic association of plant root and fungus.

N

Natural selection: Differential success in the reproduction of different phenotypes resulting from the interaction of organisms with their environment.

Net primary production: The total energy accumulated by plants during photosynthesis (gross primary production minus respiration).

Nitrification: The conversion of nitrites to nitrates, usually by microorganisms. The term is commonly used to describe the process of conversion of ammonium ions via nitrites to nitrates.

Nitrogen fixation: The assimilation of atmospheric nitrogen by certain prokaryotes into nitrogenous compounds that can be directly used by plants. The conversion of gaseous nitrogen (N_2) into more complex molecules. The process is used industrially to produce nitrogen fertilizers.

Nutrient cycling: The transformation of chemical elements from inorganic form in the environment to organic form in organisms and, via decomposition, back to inorganic form.

O

Oceanic zone: The region of water lying over deep areas beyond the continental shelf.

Oligotrophic lake: A nutrient-poor, clear, deep lake with minimum phytoplankton.

Omnivore: A heterotrophic animal that consumes both meat and plant material.

Operant conditioning: A type of associative learning in which an animal learns to associate one of its own behaviors with a reward or punishment and then tends to repeat or avoid that behavior. Also called trial-and-error learning.

Opportunistic species: One that is capable of exploiting spasmodically occurring environments.

Osmoregulation: The control of water balance in organisms living in hypertonic, hypotonic, or terrestrial environments. Regulation of the salt concentration in cells and body fluids.

Osmosis: Diffusion of water across a semi-permeable membrane.

Overnourishment: A diet that is chronically excessive in calories.

P

Parasite: An organism that obtains its nutrients from one or a very few host individuals causing harm.

Parasitism: A symbiotic relationship in which the symbiont (parasite) benefits at the expense of the host by living either within the host (as an endoparasite) or outside the host (as an ectoparasite).

Pathogen: A microorganism or virus that causes disease.

Pelagic zone: The area of the ocean past the continental shelf, with areas of open water often reaching to very great depths.

Perennial: A plant that lives for many years.

pH: The negative logarithm of the hydrogen ion concentration of a solution (expressed as moles per liter). Neutral pH is 7, values less than 7 are acidic, and those greater than 7 are basic.

Phenotype: A visible, or otherwise measurable, physical or biochemical characteristic of an organism, resulting from the interaction between the genotype and the environment.

Pheromones: Chemicals released, usually in minute amounts, by one animal, that are detected by, and act as a signal to other members of the same species.

Photoautotroph: An organism that harnesses light energy to drive the synthesis of organic compounds from carbon dioxide.

Photoperiodism: A physiological response to day length, such as flowering in plants.

Photosynthesis: The biological process that captures light energy and transforms it into the chemical energy of organic molecules (such as carbohydrates), which are manufactured from carbon dioxide and water; performed by plants, algae, and certain bacteria.

Phytoplankton: Algae and photosynthetic bacteria that drift passively in the pelagic zone of an aquatic environment.

Placental mammal: Mammals which develop a persistent placenta, i.e. all mammals other than marsupials and monotremes.

Poikilotherm: An organism whose body temperature is strongly correlated with that of its external environment.

Polymorphism: The existence within a species or population of different forms of individuals, beyond those that are the result simply of recurrent mutation.

Population: A group of individuals of one species in an area, though the size and nature of the area is defined, often arbitrarily, for the purposes of the study being undertaken.

Population density: The numbers in a population per unit area.

Population ecology: The study of how members of a population interact with their environment, focusing on factors that influence population density and growth.

Population fluctuations: Variations over time in the size of a population.

Predation: The consumption of one organism, in whole or in part, by another, where the consumed organism is alive when the consumer first attacks it.

Predator: An organism that consumes other organisms, divisible into true predators, grazers, parasites and parasitoids.

Prey: An individual liable to be, or actually, consumed (and hence killed) by a predator.

Primary producer: An autotroph, which collectively make up the trophic level of an ecosystem that ultimately supports all other levels; usually a photosynthetic organism.

Primary productivity: The rate at which biomass is produced per unit area by plants.

Producers: Organisms that make organic food molecules from CO_2, H_2O, and other inorganic raw materials: a plant, alga, or autotrophic bacterium.

Productivity: The rate at which biomass is produced per unit area by any class of organisms.

Primary succession: A type of ecological succession that occurs in a virtually lifeless area, where there were originally no organisms and where soil has not yet formed.

R

Radiation: Energy emitted from the unstable nuclei of radioactive isotopes.

r selection: Selection of life-history traits which promote an ability to multiply rapidly in numbers - the traits being, broadly, small size, precocious reproduction, semelparity, a large reproductive allocation and the production of many but small offspring.

Random distribution: Lacking pattern or order. The result of (or indistinguishable from the consequence of) chance events.

Reflex: An automatic reaction to a stimulus, mediated by the spinal cord or lower brain.

Reproduction: The production of new individuals.

Ruminant: Animals, such as cows that chew the cud and have complex stomachs containing microorganisms that break down the cellulose in plant material.

S

Saprophyte: An organism that acts as a decomposer by absorbing nutrients from dead organic matter. An organism that carries out external digestion of non-living organic matter and absorbs the products across the plasma membrane of its cells (e.g. fungi).

Savanna: A tropical grassland biome with scattered individual trees, large herbivores.

Secondary succession: A type of succession that occurs where an existing community has been cleared by some disturbance that leaves the soil intact.

Sessile organism: Literally a 'seated' organism. One whose position is fixed in space except during a dispersal phase, e.g. a rooted plant, barnacles and corals.

Species: A group whose members possess similar anatomical characteristics and have the ability to interbreed.

Species diversity: An index of community diversity that takes into account both species richness and the relative abundance of species.

Species richness: The number of species present in a community.

Steppe: Treeless plains of southeastern Europe and Siberia.

Succession: The non-seasonal, directional and continuous pattern of colonization and extinction on a site by populations.

Succulents: Plants with fleshy or juicy tissues with high water content characteristic of desert and saline environments.

Survivorship: The probability of a representative newly born individual surviving to various ages.

Survivorship curve: A plot of the declining size of a cohort, or presumed cohort, as the individuals die, usually with time on the horizontal axis and $\log_{10} lx$ on the vertical axis (where lx is the proportion of the original cohort still alive).

Symbiosis: An ecological relationship between organisms of two different species that live together in direct contact.

T

Taiga: The coniferous or boreal forest biome, characterized by considerable snow, harsh winters, short summers, and evergreen trees. The coniferous forest that extends across much of North America and Eurasia bounded by tundra to the North and by steppe to the south.

Taxonomy: The study of the rules, principles and practice of classifying living organisms.

Temperate deciduous forest: A biome located throughout midlatitude regions where there is sufficient moisture to support the growth of large, broadleaf deciduous trees.

Territoriality: The establishment by an animal or animals of an area from which other individuals are partially or totally excluded.

Tertiary consumer: A member of a trophic level of an ecosystem consisting of carnivores that eat mainly other carnivores.

Thigmotropism: The directional growth of a plant in relation to touch.

Threatened species: Species that is likely to become endangered in the foreseeable future throughout all or a significant portion of its range.

Torpor: In animals, a physiological state that conserves energy by slowing down the heart and respiratory systems.

Transpiration: The evaporation of water from a plant surface.

Trophic level: Position in the food chain assessed by the number of energy-transfer steps to reach that level.

Trophic structure: The organization of a community described in terms of energy flow through its various trophic levels.

Tundra: The biome that occurs around the Arctic circle, characterized by lichens, mosses, sedges and dwarf trees.

Turbulence: Fluid flow in which the motion at any point varies rapidly in direction and magnitude.

Turgor: The distention of living tissue due to internal pressures.

U-W-X-Z

Undernourishment: A diet that is chronically deficient in calories.

Uniform: Describing a dispersion pattern in which individuals are evenly distributed.

Variation: Differences between members of the same species.

Wavelength: The distance from one wave peak to the next; the energy of electromagnetic radiation is inversely proportional to its wavelength.

Wild type: An individual with the normal phenotype.

Xerophytes: Plants adapted to arid climates.

Zero population growth: A period of stability in population size when the per capita birth rates and death rates are equal.

Zygote: A fertilized egg. A diploid cell resulting from fertilization of an egg by a sperm cell.

REFERENCE

1. Akben F.; Sungur N. Çevre ve İnsan. Ankara: Gün Yayýncýlýk, 1994
2. Arms K.; Camp P. S. Biology A Jounery Into Life. New York: Saunders College Publishing, 1991
3. Baþaran, A. Týbbi Biyoloji. Üçüncü baský. Eskiþehir: Hünkar Ofset, 1992
4. Baþoðlu, M; Öktem, N. Zoofizyoloji Praktikumu. İzmir: Ege Üniversitesi Fen Fakültesi, 1984
5. Lewis R. Life third edition. WBC. McGraw-Hill. 1998
6. Beckett, B.S. Illustrated Human and Social Biology. 11th Edition. London: Oxford University Press,1995
7. Arpaci O. The introduction to Biology. Istanbul: Zambak publication, 2003
8. Brum, G.D.; McKane, L.K. Biology: Exploring Life U.S.A: John Wiley & Sons, 1989
9. Mauseth. Botany. Second editions. Sounders college publishing, 1995.
10. Cecie Starr. Biology Concepts and Applications. Second Edition. U.S.A: Wadworth, 1994
11. Ceren M. Biyoloji. İstanbul: Uður Yayýnlarý, 1997
12. Colinvaux P. Ecology 2. U.S.A: John Wiley & Sons Inc., 1993
13. Demirsoy, A. Yaþamýn Temel Kurallarý, Cilt-1/Kýsým 1-2. Ankara: Metaksan Kaðýt Karton Üretim Tesisleri, 1985
14. Doðan M.; Korkmaz N. Çevre ve İnsan. Ankara: Secan Yayýncýlýk Ltd. Þti., 1995
15. Kenci B. Dogan M. Arpaci O. Zoology. Istanbul: Zambak publication 2004
16. Dorit R. L.; Walker W. F.; Barnes R. D. Zoology. U.S.A: Saunders College Publishing, 1991
17. Campbell. Reece. Biology, sixth edition. USA, Benjamin Cummings publications 2002.
18. Arpaci O. Ozet M. Heather J. E. Biology 2. Istanbul: Zambak publication 2000
19. Guyton, A. Fizyoloji, Cilt 1-2-3. Ankara: Güven Kitabevi Yayýnlarý (Tercüme), 1978
20. Arpaci O. Ozet M. Heather J. E. Biology 3. Istanbul: Zambak publication 2000
21. Hopson J. L.; Wessels N. K. Essential of Biology. U.S.A: McGraw-Hill, Inc, 1990
22. Texas science grade 7. New York, USA. McGraw-Hill. 2002
23. Karamanoðlu, K. Genel Botanik. İstanbul: Çaðlayan Kitabevi, 1973
24. Kenanoðlu S.; Dinçtürk S. Botanik-Bitki Sistematiði. Ankara: Yaygýn Yükseköðretim Kurumu, 1978
25. Kocataþ A. Ekoloji (Çevre Biyolojisi). Bornova: Ege Üniversitesi Matbaasý, 1992
26. Mackean D.; Jones, B. Human and Social Biology. 5th Edition. London: John Murray Ltd, 1993
27. Arpaci O. Ecology. Istanbul: Zambak publication 2003
28. Solomon E. P.; Berg R. L.; Martin D. W. Biology fifth edition. USA: Sounders college publishing, 1999.
29. Salisbury, F.W.; Ross, C.W. Plant Physiology.4th Edition. California: Wadworth, 1992
30. Þiþli M. N. Çevre Bilim-Ekoloji. Ankara: Yeni Fersa Matbaacýlýk, 1996
31. Tanyolaç, J; Tanyolaç, T. Genel Zooloji. Ankara: Hatipoðlu Yayýnlarý, 1990
32. Toole, A.G.;Toole, S.M. A-level Biology. London: Charles Letts Books Ltd, 1982
33. Ünsal, N. Genel Biyoloji Laboratuarý (Botanik). İstanbul: İstanbul Üniversitesi Yayýnlarý, 1990
34. Villee C.; Solomon E. P.; Martin D. W.; Linda R. B. Biology. Fourth Edition. U.S.A: Saunders College Publishing, 1996
35. Wallace R. A. Biology (The World of Life). Fifth Edition. U.S.A: Harper Collins, 1990

Printed in Great Britain
by Amazon